The Radio Spectrum

The Radio Spectrum

Managing a Strategic Resource

Edited by
Jean-Marc Chaduc
Gérard Pogorel

Series Editor
Pierre-Noël Favennec

Part of this book adapted from "La gestion des fréquences" published in France in 2005 by Hermes Science/Lavoisier
First published in Great Britain and the United States in 2008 by ISTE Ltd and John Wiley & Sons, Inc.

ISTE Ltd
6 Fitzroy Square
London W1T 5DX
UK

www.iste.co.uk

John Wiley & Sons, Inc.
111 River Street
Hoboken, NJ 07030
USA

www.wiley.com

© ISTE Ltd, 2008
© GET et Lavoisier, 2005

Library of Congress Cataloging-in-Publication Data

The radio spectrum : managing a strategic resource / edited by Jean-Marc Chaduc, Gérard Pogorel.
    p. cm.
"Part of this book adapted from "La gestion des fréquences" published in France in 2005."
Includes bibliographical references and index.
ISBN: 978-1-84821-006-6
1. Radio frequency allocation--Management. I. Chaduc, Jean-Marc. II. Pogorel, Gérard.
TK6553.R3353 2008
384.54'524--dc22
                                                                            2007026597

British Library Cataloguing-in-Publication Data
A CIP record for this book is available from the British Library
ISBN: 978-1-84821-006-6

Printed and bound in Great Britain by Antony Rowe Ltd, Chippenham, Wiltshire.

FSC
Mixed Sources
Product group from well-managed
forests and other controlled sources
Cert no. SGS-COC-2953
www.fsc.org
© 1996 Forest Stewardship Council

Table of Contents

Acknowledgements and Credits

Some figures and pictures in this book are published courtesy of external bodies to whom we would like to express our thanks.

The extracts from the Radio Regulations are published by authorization of the ITU as publishing rights owner. The choice of these documents and their content, which have been adapted under the full responsibility of the authors, are only given for illustration and do not imply any responsibility of the ITU. The original Radio Regulations can be obtained from the ITU at the following address:

Union internationale des telecommunications
Secrétariat général. Service des ventes et marketing.
Place des Nations. CH-1211 GENEVA 20 (Switzerland)
Telephone (English) +41 22 730 61 41, (French) +41 22 730 61 42
Fax +41 22 730 51 94
email: sales@itu.int
internet: www.itu.int/publications

Most pictures have been kindly provided by the French Agence Nationale des Fréquences, some of them made by A. Gonin and ANFR agents. Documents on broadcasting have been provided by the company Télédiffusion de France (TDF). Significant input for this book and many illustrations have been provided by the SPORT VIEWS (www.sportviews.org) research project.

Introduction

How Can Frequencies be Managed?

Everywhere, all the time, we are bathing in a "field" of radio waves. The whole universe is covered by these waves. It maintains the everlasting trace of the initial "big bang" as a very weak radio noise which recreates a picture of the cosmos, at its very beginning for us.

Since discovering the laws of electromagnetism, during the 19th century, which are described by the set of Maxwell's equations, and inventing technical devices to produce and use these waves predicted by theory, man has added his own man-made waves to the natural ones.

Radio waves are electromagnetic waves whose frequency is low compared to others such as optical waves, X-rays or Gamma rays.

The radioelectric waves can be imagined as a shock, a vibration of space which spreads in all directions from a transmitter, traveling at the speed of light. These waves meet obstacles which absorb or deflect them and they weaken as the distance grows. After having traveled some way, they progressively mix with the radioelectric noise which exists all around, being created by innumerable wave sources, natural or man-made. Progressively, more and more sophisticated techniques are required to extract from this general noise a particular radio signal which weakens when propagated from the transmitter. If a radio link is established between a transmitting and a receiving station, the receiver must be installed in a place where, according to its sensitivity, it can separate the useful signal from the noise and process it.

Engineers have found ways to shape the completely immaterial but completely real radio waves and map them in the three dimensions of space. If the exact location of each transmitter is known with its radio characteristics, if the obstacles met by the waves are evaluated, if the propagation conditions are assessed, we can design a map of the radioelectric field, at every place, often exactly and at least in a statistical form.

Moreover, if the radioelectric noise can be described and modeled, if well defined receivers are installed, the useful range of any particular radio system can be calculated and designed. It is important to note that a radio system (transmitter, propagation medium, receivers) should be seen as a whole. Any part cannot be considered without reference to others. In fact, radio engineers must generally manage whole systems rather than transmitters alone. Their objective is to implement everywhere as many radio systems as possible, which are needed by our modern societies, without interferences between them.

The management of the radio "natural resource", which is usually called "spectrum" or more simply "frequencies", to insure a variety of practical services on the field throughout the world, is the subject of this book. It considers all possible kinds of instruments: technical, administrative, legal, economical, which may be used. However, it will not deal with radio system engineering, i.e. how to optimize the use of a specific radio resource for a given radio system, on a definite area, in conformance with regulations. As an example, the mobile radio systems or satellite system engineering are two specific subjects which might be the matter of dedicated books. A mobile operator should implement its radio stations in such a way to optimize their coverage, satisfy the customer traffic, lower its investment costs and so on. Specific tools are necessary for this.

Only general tools will be considered here which help to optimize and share the whole radio spectrum resource to satisfy the different needs, taking into account the huge variety of users and uses. Spectrum management, as far as we understand it, does limit itself to rules which help this sharing. It is a science for spectrum regulators and administrators rather than a science for operators which will be outlined further.

It could be useful to mention that this word "spectrum" has several different, albeit related, meanings. The one which will be mainly considered here is the global radio resource, available anywhere and anytime and which has to be organized for its best use. However, the same word may mean other things: as an example, a particular radio signal has an intimate structure, described by a frequency "spectrum" which occupies a frequency band. It is not difficult to make the distinction between these different meanings according to the context.

Among the natural forces, the fundamental interactions which are at work in the matter, those which are carried by radio waves are the "lightest". We could say that their influence stays on the surface of things. The energy quantum w (w = h.υ, according to Planck's formula with h being Planck's constant and υ the wave frequency) is too weak to disturb the atoms and affect the electronic cloud structure around the nucleus. Thus, radio waves are non-ionizing, having no influence on neutral atoms, thus being different from other electromagnetic waves with a higher frequency, such as optical waves. This is truly important when considering, for example, the effects of waves on health.

There has been, in the past, a long and hard debate about the intimate nature of light: is it a wave or a stream of particles? Quantum theory has given an answer. Every electromagnetic signal can be simultaneously described as a wave and a stream of particles and can be observed as one or the other according to the experimental context. However, practically, radio signals only appear as waves. Spectrum management can ignore their quantum aspect.

Radio waves interact with electrons, ions or molecules having an electric polarity. In radiocommunications, the interaction of waves with ions or polar molecules is a rather specific phenomenon which is only seen in some circumstances such as reflection on the Earth's ionosphere or transmission through the atmosphere. In such cases, only waves at specific frequencies are generally concerned. On the other hand, the interaction of waves with electrons, particularly electrons which move freely in metallic mediums, is a fundament of radio techniques, typically the basis of antenna properties. The interaction between electric currents and radio waves at the vicinity of an antenna gives shape to the wavefront which is transmitted. It also determines the strength of the electric current which is inducted in an antenna by a particular incident wavefront. Antenna theory is the basis of radiocommunications.

Actually, the very weak interaction between ordinary radio waves and matter helps to consider the radio spectrum as a rather closed domain of physical phenomena which can be managed for itself without too much care for other aspects of science or social life. In normal conditions, radio waves have no biological, chemical or mechanical impact. However, when used at a high power, they produce of course effects such as heating, which are used in microwave ovens, as an example. The recent debates concerning the impact of mobile phones on public health shows that the public deserves true and precise information in order not to react irrationally to the implementation of modern radio networks.

It is a fact that the rapid development of radio networks of all kinds in our society has changed public feelings about radio. Nowadays everybody has a mobile phone and radio stations are everywhere. Such major contemporary events such as man landing on the Moon, or the Gulf War, have clearly been achievements in radio techniques. The "great ears" of the secret services may be seen as a threat to privacy. Our world is becoming a radio world where waves are weaving an ever denser net around the Earth.

Positively, radio is a basic technology for building the so-called "world village". Available anywhere, at any time, capable of building links at very short distances as well as on a cosmic scale, radio is a unique tool to connect men and things without any material medium. It is also a tremendous tool for social progress. New telecommunications services are often the offspring of radio such as broadcasting in the 1920s, TV in the 1950s, satellites in the 1960s, mobile phones in the 1990s, and radiolocation in the 2000s. But new regulations and new market opportunities are also born from radio implementations. The public debate which has taken place during the last 20 years about deregulation of the information society initially focused on traditional telecommunications cable networks and on the historical telephone operator monopoly. However, radio shuffled the cards and introduced a "wireless society" which is now universally adopted and made most of the former considerations on cable network regulations obsolete. When reading again the innumerable books dealing with deregulation, written during the period 1970-1980, it may be regretted that so many pages deal with cables and so few with radio. Experts are often short-sighted. Now, radio is in the foreground and radio spectrum management is a strategic issue, much more so than cable ownership.

Spectrum management can now be seen as a major goal for telecommunications efficiency. It is necessary that this natural and public resource be utilized for the profit of as many users as possible, taking care of the largest variety of needs. Let us understand that the spectrum resource is not capacity limited. The management of frequencies, if always becoming more accurate and clever, creates ever increasing possibilities.

We are going to describe the main tools available for this management. The book is made up of two parts.

Part 1, up to Chapter 9, is more technical. It deals with the management methods based on electromagnetism laws, information theory, geographical constraints. Basic rules and procedures are described which have been progressively elaborated from the beginning of the 20th century to share the spectrum resource. These rules are backed by international legal provisions which form a radio spectrum law. Current uses of radio are considered with their recent evolutions.

Part 2, more prospectively, describes the bodies in charge of this management with their functions and activities, with reference to the main questions they have to answer. New opportunities to improve the management of spectrum are presented.

Of course, this matter is rapidly changing. However, the general movement has now existed for more than a century. It can be judged that spectrum management has steadily progressed at a continuous pace during this time with some important changes at crucial periods. This is a reason to be confident for the necessary evolutions in the future.

Part 1

The Basis of Spectrum Management

Chapter 1

A Bit of History, Physics and Mathematics

Electricity and magnetism have been known about for a long time. As long ago as Antiquity, men wondered about the properties of magnets. 18th century amateurs were fascinated by these strange forces which acted at a distance and were recognized as being probably linked together. 19th century scientists began their theoretical approach. Ampère and Oersted demonstrated the reciprocal influence of magnets and electric currents through magnetic "fields" which carried forces capable of distant actions without any intermediate material medium. Notably, the induction phenomenon was discovered: when varying in a circuit, an electric current can create a similar electric current in a distant coil. Induction clearly opened a new chapter in the history of physics and offered new technical opportunities to communicate at a distance. However, we had to wait several years more for completely satisfactory synthesis to open the way to radio, and so enjoy the fruits of this hard work.

In fact, radio was born in the 20th century. Even with the support of Maxwell's fundamental theory, published by 1860-1875, whose beautiful equations showed the evidence of electromagnetic waves traveling at the speed of light, the demonstration of making these waves carry a useful signal had to wait for Marconi, in around 1895.

The distribution of waves in space around a radio transmitter is called the radioelectric field.

After Maxwell, different experimenters came, among them Hertz, who demonstrated in 1887 that this field was clearly wave-shaped, showing all the typical features of a vibratory phenomenon. He described stationary vibratory fields with stable maximums and minimums in space and assessed the possibility of realizing tuned circuits which generate electric oscillations in resonance with an incident wave at the same frequency. The works by Hertz and others paved the way for technical devices which can efficiently transmit and receive radio waves:

– the generator (emitter) which produces electrical oscillations at a given frequency;

– the tuned antenna which radiates and collects the electromagnetic waves in space;

– the receiver which chooses with an appropriate selectivity the signals to be processed, in order to recover any useful message.

In the same manner as an organ pipe or a piano string resonates at an acoustical wave whose frequency fits its tuning, an antenna resonates at an electromagnetic wave.

These first theoretical and practical approaches may help in understanding some fundamental technical bases for radio spectrum management.

1.1. Waves

An electromagnetic field may be pictured as a wave which is propagated at light speed "c" independently of any other characteristics (c = 300,000 km/s, approximately, in a vacuum or in the atmosphere). Light is a particular electromagnetic wave with a very high frequency, while radio waves are electromagnetic waves with the lowest frequencies.

At any place in space an electromagnetic field combines two fields, an electric field E and a magnetic field H, which are fully proportional at a sufficient distance from the transmitting antenna. Both basic fields are "vectors", which means that they have an intensity (amplitude) and a direction in space. E and H are perpendicular to each other and are included in the wavefront which is propagated perpendicularly along the direction of the "propagation vector", P, also called Poynting's vector. The amplitude of this P vector represents the electromagnetic power flux density which is carried by the wave.

The instantaneous value of E (or H) at a place changes very rapidly according to the wave rhythm (the frequency) as a sine function:

$$E = E_0 \sin (2\pi f . t + \varphi_0)$$

where f is the frequency, t the time, Eo the amplitude (or intensity) and φo a constant depending on the place where the wave is considered.

The electric field intensity unit is the volt per meter (V/m). This same unit is used to measure the radio field intensity.

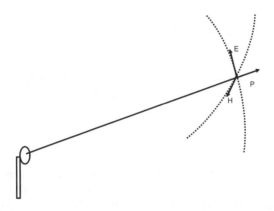

Figure 1.1. *Electromagnetic field components. The wavefront and Poynting's vector. The electromagnetic field comprises an electric field E and a magnetic field H which are proportional and perpendicular to each other. They are included in the wavefront which is propagated from the transmitting antenna. The propagation direction is represented by vector P, called Poynting's vector, the amplitude of which represents the radioelectric power flux density (PFD). This PFD determines the radioelectric power which a receiving antenna, at this place, may catch*

A wave carries electromagnetic energy, as a power flux spread on the wavefront. As mentioned above, at some distance from the transmitting antenna, the power flux density (PFD) is represented by Poynting's vector. It is proportional to the squared amplitude of the electric field:

$$PFD = k \ (Eo)^2$$

The radio power which passes through a surface S, which is parallel to an incident wavefront, is

$$W = PFD \ . \ S$$

Thus, the PFD is expressed in watts per square meter (W/m²).

If the surface is not parallel with the wavefront, the power which crosses it is given by:

$$W = PFD . S . \cos \theta,$$

θ being the angle between the propagation vector and the perpendicular line of the surface.

Power is a scalar quantity, i.e. a simple number without direction, which is the opposite of the field which is space-directed, being a vector.

$$W = k (Eo)^2 . S \tag{1.1}$$

Being a scalar, radio power or energy which is captured by any equipment, at any point in space, can be calculated simply by adding the different power contributions from the different radio waves. On the contrary, the global electromagnetic field, as a vector, should be calculated using a vectorial addition of elementary fields which is much more complicated.

The vectorial addition of fields must be used for distributed radio system engineering when the contribution of every radio source is precisely known. As an example the radiation pattern of a complex array of elementary antennae should be calculated by the vectorial addition of the fields created by the different antenna elements. The most spectacular case is probably the huge radio interferometers used for radio astronomy.

A number of antennae, especially those used for TV broadcasting or in mobile base stations are built from a grid of elementary dipoles, a type of rod, placed in parallel, side by side, in a vertical plan. This arrangement produces a broad coverage in the horizontal plane, in front of the station, but it also "pinches" the vertical radiation pattern and concentrates the radio power flux on the ground to strengthen the field towards the customer receivers.

The scalar addition of energies is less informative. It can be used when the exact geometric components of the local radio field are not known. It is typically useful to evaluate the global radio noise which does exist somewhere or the energy absorbed by a human body at a place.

Mention should also be made of another significant feature describing radio waves, when managing the spectrum resource, which is their polarization. Polarization is the electric field direction in the wavefront. A wave may be linearly polarized when the electric field keeps the same direction during the wave propagation. It may be circularly polarized when the field direction rotates at the same frequency as it vibrates. These direction properties may be usefully controlled for a more efficient use of different waves at the same frequency but with perpendicular polarizations, because it is easy to discriminate two such waves. This opportunity is not at the very heart of the spectrum management techniques which will be described later, but is a valuable extra to improve the efficiency of more general and powerful methods.

A vertical antenna, for example, radiates a vertically polarized wave. Thus a vertical antenna will normally be necessary to catch it. On the contrary, a horizontal antenna will catch practically no signal if the propagation takes place in free space. In fact it should be noted that the initial wave polarization is only maintained for waves traveling in such a free space, and any reflection may modify it. TV is often broadcast with horizontally polarized waves. This fact does explain why the Yagi receiving antennae on house roofs represent a horizontal grid.

As with any vibration, a radio wave is mainly described by its frequency "f" which is the number of oscillations per second and which is expressed in hertz ("cycles", in the past). One hertz means one oscillation per second. The radio waves occupy a frequency range from a few kilohertz to several hundred gigahertz. This broad domain, called the radio spectrum, cannot be described on a unique scale and has to be more easily referred to in different frequency units:

– a kilohertz (kHz), 1,000 Hz;

– a megahertz (MHz), 1,000,000 Hz;

– a gigahertz (GHz), 1,000,000,000 Hz.

The border between the radio spectrum and the infrared optical spectrum is unclear. The Radio Regulations have placed this point at 3,000 GHz. This figure has to be compared with the frequency of common optical waves, typically 0.6 µm of wavelength, which is 500,000 GHz. However, a more practical upper value for usable radio frequencies could be 300 GHz, the limit of millimetric waves (EHF).

The ratio between 300 GHz and 10 kHz is 30,000,000, or 25 octaves, to be compared, for example, with the visible light spectrum which covers approximately one octave. It is thus understandable that radio wave properties vary considerably throughout the spectrum and imply different management practices from low frequency waves to microwaves, even if the theoretical basis remains unchanged.

A fundamental property of a vacuum (or ordinary atmospheric space), is linearity. This means that radio waves which are propagated in such a space never interfere or mix together. We can add as many waves as wanted or necessary, at the same place, without any influence between them. Each one acts as if it were alone. The whole spectrum management is based on this principle that it is possible to recognize and process the different waves using appropriate technical devices, notably due to their respective frequencies. For that purpose, the receiver "selectivity" is a major parameter for its performance. However, waves can also be distinguished by some other parameters describing the vectorial aspects of the radio field: the propagation Poynting's vector and the polarization. Thus, the antenna "directivity" and polarization should be considered.

Based on the receiver selectivity, it may be judged obvious that one of the basic practices for spectrum management is simply to share it between a number of frequency bands which are small parts of the huge global spectrum. Each spectrum band or frequency slot should have homogenous and uniform properties and could be easily distinguished (or "filtered") from the neighboring bands. In doing so, the job is made easier since we can ignore the signals which are situated in other remote spectrum parts.

As an example, a radio set can be designed to receive the FM band, 87.5 to 108 MHz. It will be completely insensitive to a TV signal in the 470-862 MHz band or to a mobile phone wave in the 925-960 MHz band. Thus, these different bands can be managed independently.

The other scale with which we can describe the waves is their wavelength, i.e. the vibration pitch, the distance in space between two points which are, at the same moment, in the same vibration state. The wavelength is expressed by the Greek letter λ and is directly related to frequency by the equation:

$$\lambda = c/f \qquad\qquad\qquad [1.2]$$

where c is the speed of light and f the frequency.

The wavelength and frequency are thus equivalent for characterizing a wave which travels in a vacuum or in the atmosphere. A basic calculation shows that the wavelength varies from 30 kilometers for a 10 kHz wave to one millimeter for a 300 GHz wave. It is less than one micrometer for an optical wave.

Wavelength is the parameter which gives the most appropriate representation of the waves relative to mechanical and geographical considerations, such as the technology of antennae and circuits or the radio station locations. It was the first parameter that Hertz measured when trying to experiment with radio waves: two maximums or minimums of stationary wave intensity are half a wavelength apart.

Some waves are used and managed on a planet scale and may require technologies with kilometric dimensions. Others apply to radio links at a distance of meters and work with tiny electronic circuits. However, any systematic approach would be inappropriate. Very small radio equipment can process long waves and huge installations may be dedicated to listening to very short wavelength signals such as large radiotelescopes.

The wavelength is a "natural" means of classifying waves:

- kilometric waves, below 300 kHz   (LF)

- hectometric waves, between 300 kHz and 3 MHz   (MF)

- decametric waves, between 3 MHz and 30 MHz   (HF)

- metric waves, between 30 MHz and 300 MHz   (VHF)

- decimetric waves, between 300 MHz and 3 GHz   (UHF)

- centimetric waves, between 3 GHz and 30 GHz   (SHF)

- millimetric waves, between 30 GHz and 300 GHz   (EHF)

At different periods more technical names or acronyms have been used for some particular bands. The Atlantic City conference, in 1947, designated the different bands as indicated above:

- LF, low frequencies, for kilometric waves;

- MF, medium frequencies, for hectometric waves, etc.

There may also be found in the technical literature, some acronyms such as L, C, X, Ku, Ka, etc. bands. They are only familiar short names to be used for specific purposes. As an example they are commonly used in satellite communication engineering to designate the different frequency domains which are allocated to this service by the Radio Regulations:

- L band: 1.5 GHz;

- C band: 4 and 6 GHz;

- X band: 7 and 8 GHz;

- Ku band: 11, 12 and 14 GHz;

- Ka band: 20-30 GHz.

Let us return to Maxwell and Hertz.

Wavelength and geometry are linked. When Hertz was designing his first experimental equipment, he worked with human-sized devices. The oscillating circuits which he built in his laboratory had dimensions of around a meter, thus he observed waves whose frequency was about a few hundred megahertz. When pioneers achieved the first long distance transmissions, they used "long waves", kilometric waves, with wire antennae covering large areas in the country. Today, when looking at the Yagi-type TV roof antennae, we can roughly evaluate their frequency band. The distance between antenna rods is typically sized between a quarter and half a wavelength and the rod length is about half a wavelength. Thus, a common TV receiving antenna, being tuned to about 600 MHz, has rods about 25 cm long and 12.5 to 25 cm distant.

1.2. Propagation

It was rapidly observed that waves fade as the distance from the transmitter increases. The first experiments with radio took place at very short distances, typically a few tens of meters. Beyond this, nothing was detected. The reason for this is that in free space, the power flux density carried by the wavefront decreases as the square of the distance.

This can be easily understood when considering that the transmitted radio power W which is generated in the transmitting station and radiated by the antenna spreads itself on a sphere, whose radius is the distance d. Thus, the power flux density (PFD) which will meet a receiving antenna decreases as the sphere surface which is proportional to the squared distance. The radio power captured by this antenna decreases in the same way.

$$PFD = W/S = W/4\pi d^2 \qquad\qquad [1.3]$$

This simple argument about the PFD decrease with distance, based on a spherical wave transmitted from an "isotropic" antenna which radiates in a vacuum, does apply in fact to any antenna, whatever its radiation pattern.

However, the steady decrease of PFD is not a rule in ordinary conditions. Some physical phenomena make the wave propagation differ from what it would be in the vacuum space.

As an example, decametric wave propagation was soon found to be much more efficient thanks to reflections on the ionosphere belts which surround the Earth. From the beginning, this advantage has often been used for long distance radio transmissions. It was, in fact, the only means of intercontinental radio communication before satellites. Other circumstances are also beneficial such as a "guidance" phenomenon which sometimes occurs when waves are propagated over the sea and are "guided" by a duct produced by atmospheric slices at different temperatures. Some other positive effects are obtained from diffracting obstacles. In all these contexts, exceptional ranges are observed, much longer than expected from the propagation in a vacuum.

On the contrary, in many circumstances, many different phenomena attenuate the waves at a shorter distance. Physical obstacles (masks) or erratic reflections can produce blank zones where nearly no signal can be detected, even not far from the transmitter. The atmosphere, itself, is not fully transparent, due to rain for example. It may even be a permanent obstacle to some microwaves.

Radio wave propagation is very complex, especially for low frequency waves, and should be taken into account by the regulation. In many cases too, the propagation losses are not constant but may vary in time due to changing path conditions. These variations may be short-term ones as encountered by mobile phones moving in a town or long-term ones, for example when solar activity impacts the ionosphere density.

Let us only remember now that a fundamental parameter to be considered when managing and regulating the spectrum is propagation loss or propagation attenuation Ap. It is one of the main natural mechanisms which should be modeled to control the radio resource and to keep it confined to a service area with the objective of re-using it at a specific distance for other purposes.

1.3. Directivity

An interesting property of electromagnetic waves is their directivity. Everybody knows from common experience that light can be directed by lenses and mirrors and that large amounts of light can be collected by some optical devices such as telescopes. Antennae can produce the same effect with radio waves.

Some antennae look like optical systems such as the parabolic antennae which are used with microwaves, whose frequencies are typically higher than 3 GHz. Parabolic roof antennae for direct satellite TV reception are a familiar example. Other antennae, similar to those used by Hertz, are simple wires or rods whose length varies according to the band considered. Car quarter-wave antennae are an example of this. Many other shapes are found: the roof Yagi rakes, helices, horns, etc. Each has its own properties.

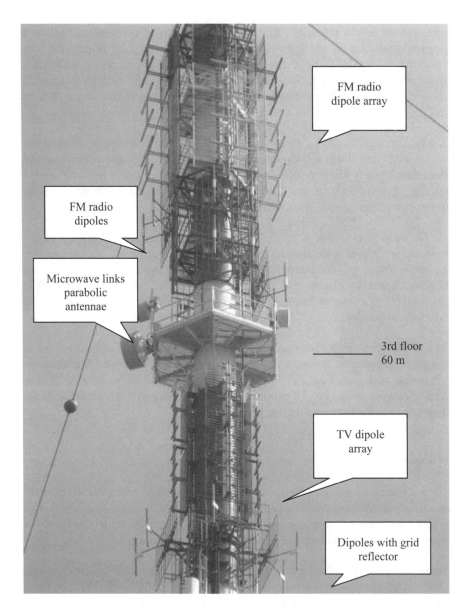

a) *Different radio stations; third floor of a radio tower*

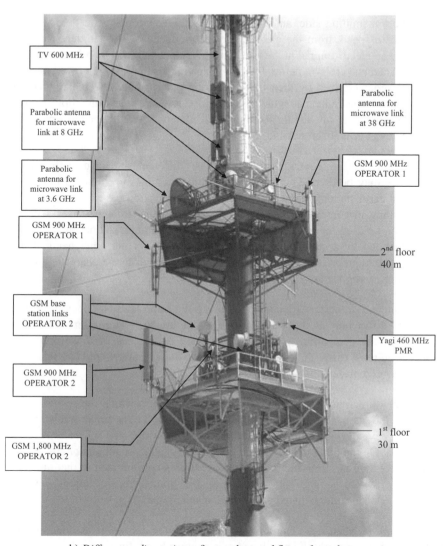

TV 600 MHz

Parabolic antenna
for microwave
link at 8 GHz

Parabolic
antenna for
microwave link
at 3.6 GHz

GSM 900 MHz
OPERATOR 1

GSM base
station links
OPERATOR 2

GSM 900 MHz
OPERATOR 2

GSM 1,800 MHz
OPERATOR 2

Parabolic
antenna for
microwave link
at 38 GHz

GSM 900 MHz
OPERATOR 1

2nd floor
40 m

Yagi 460 MHz
PMR

1st floor
30 m

b) *Different radio stations; first and second floor of a radio tower*

Figures 1.2a and b. *TV towers gather many radio systems for different services and operators. Various types of antennae may be seen here: parabolic dishes, dipoles, dipole arrays, Yagi, etc. The systems are operating in very different frequency bands, from about 100 MHz (FM) to 40 GHz (microwave radio links). The wave polarizations are vertical or horizontal. It may be noted that, in such TV towers, many microwave links are used to connect the TV broadcasting station or the radiotelephone base stations to distant infrastructures, such as a studio or switching center*

On the transmitting side, an antenna creates the electromagnetic field which is propagated in space, from the electric current produced by the emitter. This field will induce an electric signal in the receiving antenna which will be processed by the receiver. Antennae are symmetric devices, which means that they are reversible, and are mainly characterized by their directivity and polarization. Of course, there are also omnidirectional antennae which are designed to work in any direction. Another quality to consider is whether an antenna is selective or not: is it tuned to a particular frequency band or is it "broadband"?

According to the context and objectives, different antennae with various possibilities may be used. The antenna on a broadcasting tower should not be the same as an earth station antenna for satellite communications. A radiotelescope has nearly nothing in common with a mobile phone antenna. In some cases a unique antenna can be used for transmission and reception in a radio station, the two paths being separated by a component called a "duplexer". On the other hand, it may sometimes be appropriate to use different antennae with different characteristics for the two transmission sides.

The antenna properties are calculated using the same laws as optics: reflection, refraction, diffraction. If the antenna dimensions are very large compared to the wavelength, the geometric optics laws may give a fair idea of those properties. However, in most cases, the complexity of the diffraction laws should be taken into account to get representative results.

Antenna directivity is described by a radiation pattern which represents, in any direction, the gain from the "isotropic" reference. The isotropic radiator is a theoretical device which radiates the radio power uniformly on a sphere centered on the antenna. On the other hand, when there is some "gain" in a direction, the field is greater there than the isotropic one: the power flux is concentrated in that direction. Of course, if there is "gain" in a particular direction, there are deficits (negative gain) in others since the radiated power remains the same.

To simplify, when it is commonly spoken about the gain of an antenna, without any more precision, only the privileged axis is considered, for nominal use. Around this axis, the gain pattern represents a "main lobe". However, it must be clear that, practically, the antenna radiates in all directions through "secondary lobes" at a much lower level than in the main lobe. This secondary radiation, which is not useful, must be restricted as much as possible, being a radioelectric "noise" which spoils the surrounding space. Of course this is only a matter of terminology and depends on the engineer's objectives. The radio "noise" is strictly identical to the "useful" wave and is propagated in the same way.

The radiation pattern is three-dimensional. However, its representation is generally a two dimension figure.

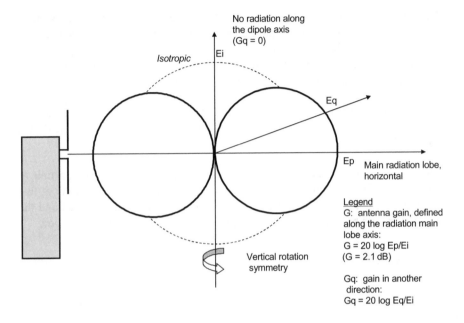

Figure 1.3. *Simplified radiation pattern of a vertical half-wave dipole. The radioelectric field amplitude as compared to the field created by an isotropic antenna is represented here in every direction of a vertical plane. It can be seen that the field radiated along the dipole is zero, which means that two dipoles placed along a same axis are perfectly isolated one from another*

It may be accepted, without any mathematical demonstration that, according to Maxwell's laws in free space, the electric field created at a distance d by an isotropic antenna which radiates a power P is:

$E = (30 . P)^{1/2}/d$, with P expressed in watts, d in meters, E in volts per meter

or $E = (30 . P . G)^{1/2}/d$, if the antenna gain is G.   [1.4]

As an example, an isotropic antenna which radiates 100 watts creates, at a distance of 10 km, in the free space, a field of 5.5 mV/m (–45 dBV/m; see section 1.4). This field will be 11 mV/m if the antenna provides a 6 dB gain.

It should be noted, from this formula, that the radio field amplitude decreases linearly with the distance and is independent of the frequency.

This linear decrease of the field intensity with distance in free space is a general property, independent of the transmitting antenna diagram, and is observed (after a distance of a few wavelengths) in any direction. This is useful for assessing the radio field radiated in the line-of-sight vicinity of a transmitting station, typically when being concerned by public health protection against radio waves.

For example, if a particular UHF or VHF transmitting station radiates on a line-of-sight location, 100 meters far, a 1 V/m field, it radiates a 0.5 V/m field at a distance of 200 meters.

The gain is a power ratio or a power flux density ratio. The common unit to express its value is the decibel (dB) which is ten times the decimal logarithm of this ratio. Between the actual PFD radiated by the antenna at a particular point and the PFDi which would be radiated by an isotropic antenna at the same place, we get:

$$G = 10 \log PFD/PFDi \tag{1.5}$$

A gain of 1 (no gain, no losses) is expressed by 0 dB (since $\log 1 = 0$). A gain of 10 is 10 dB, a gain of 100 is 20 dB, a gain of 1,000 is 30 dB and so on. A gain of 2 is commonly expressed by 3 dB (since $10 \log 2 = 3.01$).

As explained above, the local power flux density is proportional to the squared electric field amplitude [PFD= k (Eo)^2]. Thus the gain may also be represented as a ratio of electric field amplitudes E/Ei, according to the formula:

$$G = 20 \log E/Ei \tag{1.6}$$

where E is the actual field amplitude at a particular point and Ei the field amplitude which would be radiated by an isotropic antenna.

These equations are strictly equivalent since $10 \log PFD/PFDi = 10 \log (E/Ei)^2 = 20 \log E/Ei$.

Gain and directivity are two faces of the same phenomenon. Saying that an antenna has more gain is saying that it is more directive. From its own definition, an isotropic radiator has a gain of 1. Outside of the main lobe of a directive antenna, the gain is normally negative since there the radiation level is lower than it would be from an isotropic antenna (the logarithm of a number lower than 1 is negative).

The different types of antennae provide very different gains. The more gain that is required, the larger the antennae should be according to diffraction theory. For long waves, even when using huge antenna structures, the gain may be limited to a few dB with a directivity angle of some degrees. On the other hand, using microwaves with antennae whose diameter could represent 100 to 1,000 times the wavelength, gains of 40 dB or 60 dB can be obtained with directivity better than a tenth of a degree.

On the contrary, miniaturization goes against gain and directivity. Basically, the tiny mobile phones used for mobile telephony cannot provide any gain or directivity since their antenna has roughly the same dimension as wavelength. As an example, a mobile phone working at 1,800 MHz deals with waves whose length is 16.6 cm in a 10 cm long box. Thus such equipment provides a "negative gain". They radiate or catch less power than an isotropic antenna and they offer no directivity. Typically a "gain" of -10 dB is common in this case.

When an antenna is directive, it should be carefully directed, with a precision compatible with the main lobe of the radiation pattern. The radio link performance is calculated taking into account the nominal antenna gains. Thus the link is generally interrupted when any disorientation occurs.

Directivity concerns both transmitting and receiving sides. The equations are identical since the antennae are symmetric. However, calculating the gain and directivity of most antennae from their geometric dimensions is not simple. Considering the ideal structure of a radiating aperture (such as a good parabolic mirror) illuminated by a uniform field, the gain is given by:

$$G = \eta \, (\pi . \, \varphi/\lambda)^2 \qquad\qquad [1.7]$$

with η a quality factor whose value is typically between 0.5 and 0.8, φ the aperture diameter and λ the wavelength.

It can be clearly seen on such an example, that the gain (or the directivity) is determined by the ratio between the geometric size of the antenna and the wavelength used.

For the same antenna, the main lobe has an angular directivity given by α (half of the main lobe angular aperture):

$$\alpha = 4/G^{1/2} \text{ with } \alpha \text{ in radians} \qquad\qquad [1.8]$$

Let us consider a parabolic antenna with a 10 m diameter, working at 6 GHz. The wavelength is $\lambda = 0.05$ m. The basic ratio φ/λ is 200. With a quality factor $\eta = 0.8$ the antenna gain is 320,000, i.e. 55 dB. The half aperture angle of the main lobe is 7 milliradians or 0.2 degrees. Thus, such an antenna should typically be directed with a precision better than 0.1 degrees.

Practically, an abacus is used to evaluate the gain and directivity of most antenna types.

It is not often possible to give a detailed expression of the gain in any direction. Generally it is enough to consider with some detail the antenna main lobe with its nominal gain value along the axis and to evaluate the secondary lobes as being "included" or limited by a sphere in all other directions. This sphere radius is the maximum value which may be observed from the different secondary lobes. This model may be found appropriate in most cases, especially for interference calculations. However, in some cases, for example the antennae of satellite earth stations, it is necessary to get a value of the gain in the first secondary lobe apart from the axis because it may be affected by interferences with satellites in an orbital position close to the nominal one.

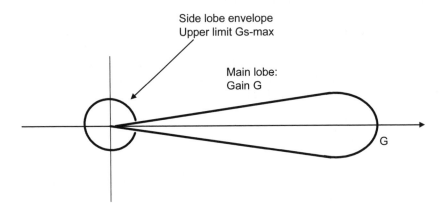

Figure 1.4. *Simplified antenna radiation pattern for coordination calculations*

1.4. Link evaluation

Now we have all the tools necessary to evaluate a global radio link, to calculate the received power from a distant transmitter, knowing the antenna characteristics and the relative situation of the two radio stations.

Received power = Transmitted power . Transmit gain . Receive gain/Propagation attenuation.

$$Pr = Pe \cdot Ge \cdot Gr/Ap \qquad\qquad [1.9]$$

This multiplicative expression is generally converted into an additive one, using logarithms

$$Pr\ (dBm) = Pe\ (dBm) + Ge\ (dB) + Gr\ (dB) - Ap\ (dB) \qquad [1.10]$$

Pr and Pe are expressed as dB relative to a reference power unit: watt, milliwatt, picowatt or others. Here dBm means dB relative to milliwatts. Ge, Gr and Ap are expressed in dB as being pure ratios.

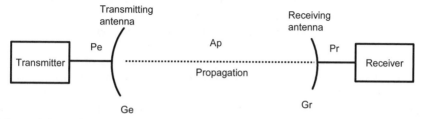

Figure 1.4a. *Link evaluation. The received power Pr may be calculated from the following parameters: transmitted power Pe, transmitting antenna gain Ge, receiving antenna gain Gr, propagation path losses Ap with Pr = Pe . Ge . Gr/Ap*

This can be expressed by a sentence: the received power equals the transmitted power, plus the transmitting and the receiving gains, minus the propagation attenuation (using a logarithmic notation, of course). However, it may be noted that any real radio installation introduces some losses on the radio and electric signals, such as feeder losses between the power amplifier and the transmitting antenna. To simplify the presentation, these losses are not mentioned here in the equations. They may be considered as included in the antenna gains which should be consequently lowered.

Note: logarithmic expressions

In radiocommunications, to ease calculations, logarithmic expressions are generally used. In such a representation a power is expressed relative to a power unit appropriate to the phenomenon scale: watt, milliwatt, microwatt, etc. and noted dBW, dBm, dBμW, etc.

Power P is expressed in dBW by P (dBW) = 10 log P (W).

The same power can be expressed in dBm: P (dBm) = 10 log P (mW).

Thus, a 10 watts power is 10 dBW or 40 dBm.

A 1 microwatt power is –60 dBW, –30 dBm or 0 dBμW.

Let us consider a 20 watts radio transmitter (13 dBW), followed by a transmitting antenna with a gain of 10 (10 dB). Suppose that the wave is received by an antenna whose receiving gain is 200 (23 dB), after having encountered propagation attenuation and various losses of 50,000,000 (77 dB). Thus the received power is Pr = 13 (dBW) + 10 (dB) + 23 (dB) – 77 (dB) = –31 dBW which is also –1 dBm or 29 dBμW.

A logarithmic expression is also often preferred for the electromagnetic field value. The units are then dBV/m, dBmV/m, dBμV/m, etc. A field intensity expressed in dBV/m is 20 log E with E in volts per meter (V/m). The same field intensity in dBμV/m is also 20 log E but with E in microvolts per meter (μV/m).

Care should be taken about these logarithmic expressions of power and field intensity. Power or power ratios are expressed by 10 times the decimal logarithm. Fields and field ratios are expressed by 20 times the decimal logarithm. This difference refers to equation [1.1] which says that power is in proportion to the squared field amplitude.

The fundamental equation [1.10] is broadly used for frequency management and various other purposes. Of course it is the most straightforward way to evaluate the capability of the nominal radio link established between two stations to work properly. But it is also used to calculate the level of interferences between different links or the noise created by distant radio sources when antenna secondary lobes are concerned.

Let us consider as an example the same 20 watts transmitter (13 dBW) as above and let us study the parasitic radiation of its antenna through a secondary lobe, whose "gain" is –8 dB. Let us suppose that at the same distance as above, this parasitic signal is captured by a lateral lobe of the receiving antenna belonging to another radio network and whose "gain" is –5 dB. The propagation attenuation is the same as above if the distance is the same, –77 db. Thus the received interference power is Ir = –77 dBW (Ir = 13 – 8 – 5 – 77).

If the foreseen equations are combined together, and supposing a free space propagation, a result is obtained:

$$Pr = Pe . Ge . Gr . (\lambda/4\pi d)^2 \qquad\qquad [1.11]$$

where the propagation attenuation appears in brackets as $(4\pi d/\lambda)^2$.

Let us consider a 150 MHz wave, whose wavelength is 2 m. In free space, the attenuation will be 100 dB after a propagation of 16 km (100 dB is 10 billions). At the same distance, a 15 GHz wave whose wavelength is 2 cm will experience a 140 dB attenuation.

Here also we can observe that geometry and wavelength are closely linked: the attenuation depends on the ratio d/λ, as the antenna gain depends on the ratio between the antenna size and the wavelength.

However, as already mentioned before, these simple formulae and magnitude orders are valid for propagation in a free space. For satellite and space communications, they fit pretty well with reality. In terrestrial communications, on the other hand, they may be altered by a number of phenomena which have to be assessed, depending on the frequency band concerned. As an example, considering a decametric wave, in the 3 to 40 MHz band, two propagation modes should be distinguished: a ground mode, with a limited range, typically several tens of kilometers, and a long distance mode by reflection on ionosphere belts at several hundred kilometers above the Earth, where the Sun ionizes the atoms. This second mode follows specific propagation laws and gives enormous radio ranges of several thousand kilometers, which direct waves cannot normally reach, because of the Earth's rotundity.

In radio regulation, two quantities are commonly used:

– the equivalent isotropic radiated power (EIRP);

– the power flux density (PFD).

EIRP represents the power which an isotropic antenna should radiate to create the same radioelectric field as an actual station along the main axis of its transmitting antenna. In fact:

$$\text{EIRP} = \text{Pe} \cdot \text{Ge} \qquad [1.12]$$

A variant of EIRP is ARP (apparent radiated power), often used for calculations of metric wave links. In the stations, half-wave dipole antennae are often installed whose length is half the wavelength. These antennae have a 1.64 gain (2.1 dB). Thus it is natural to compare the performance of any antenna not to an isotropic radiator, which is a theoretical concept, but to a half-wave dipole. It is easy to get the following equivalence:

$$\text{EIRP} = 1.64 \text{ ARP or EIRP} = \text{ARP} + 2.1 \text{ dB}$$

This EIRP concept is easily extended to evaluate the equivalent power radiated by an antenna through its secondary lobes. This secondary EIRP may be useful for interference calculations and electromagnetic compatibility.

As seen before, the PFD describes the radio power flux, measured as watts per square meter (W/m²). According to its definition, the radio power which crosses a surface S parallel to the wavefront is W = PFD .S

Sometimes it is preferred not to characterize an antenna by its gain but by its "effective area" Se. On the receiving side, this quantity expresses the capability of the antenna to capture more or less energy from the wave. Thus an antenna with an effective area Se, set parallel to the wavefront, captures a power Pr:

$$Pr = PFD . Se \hspace{5cm} [1.13]$$

We should not concern ourselves with the terminology. The "effective area" may appear more concrete than the "gain". However, no evident relationship exists between the geometric size or surface of an antenna and its "effective area". What is, as an example, the effective area of a wire antenna? As the gain, this value Se must be calculated from Maxwell's equations. However, like gain, it may be related to the actual antenna area in some cases where the geometric approximation can apply. In fact gain and effective area are equivalent: $G = 4\pi/\lambda^2 . Se$.

As an example, considering a parabolic antenna with a uniform illumination and a diameter much larger than the wavelength, the effective area is Se = η S (see [1.7]).

The PFD is commonly employed to characterize the radio flux transmitted by a satellite to illuminate a particular terrestrial area. If a satellite radiates a definite EIRP towards a region on the Earth, the PFD at that place is

$$PFD = k . EIRP/4\pi d^2 \hspace{4cm} [1.14]$$

with d being the distance between the satellite and the considered place and k an attenuation factor due to the atmospheric losses (which are generally small).

If a receiving station is there, with an effective area Se, it will capture a radio power

$$Pr = k. EIRP . Se/4\pi d^2 \hspace{4cm} [1.15]$$

PFD is also often used for space radio coordination, for example to set rules on the radiations which are radiated by a satellite to different locations or directions on the Earth. Such limits are necessary to protect different space radio networks from each other, when they share the same frequency band.

All quantities and technical notions described in this first chapter are commonly used for frequency management with two main objectives:

– to determine the service area of any particular radio system or network;

– to set the limits which any system must respect not to interfere with other ones.

These two objectives should always be considered together. Spectrum management is a permanent equilibrium between rights and duties that any radio system must comply with.

All the theoretical considerations which have been introduced above are applicable throughout the whole spectrum. However, it has been explained that a unique physical reality can be best described with different tools and quantities, according to the context:

– electric field versus power flux density;

– antenna gain versus effective area.

For great wavelengths, typically down to decimetric waves, it is mostly preferred to use electric field representations. The formulae are well suited to terrestrial mobile radio services or broadcasting services when a service area should be described where mobiles and receivers will be installed.

For short wavelengths, typically microwave systems above 3 GHz, it is preferable to work with power units (EIRP, PFD and others). They are well suited to describe satellite or fixed microwave links.

Chapter 2

Telecommunications

Electromagnetic waves are remarkable vectors in their ability to act at a distance without any material intermediate support.

However, to use their capacity for transporting information, it is not enough to simply transmit and receive waves. They are only vehicles, "carriers" according to the technical vocabulary, which should be loaded with a message. Pioneers such as Branly, Lodge, Popoff and Marconi, contributed to this achievement between 1890 and 1895.

In fact, a pure, stable radio wave, by itself, does not carry any information. Any message or "meaning" must be associated with some variation of the wave characteristics. Only these changes, being detected and measured in a receiver, may bring information.

The modification of some wave characteristics according to a message which will be carried away is called "modulation". "Demodulation" is the inverse operation where wave variations are detected and the message extracted from these variations. It was Branly's invention, in 1890, which created the first practical device capable of translating into a variable electric current the presence or absence of a radio wave. This was enough to demodulate a wave which was binary modulated with an elementary on-off message.

It is then necessary to clearly distinguish, in radiocommunications, the carrier wave which is propagated according to its own physical laws, as seen in Chapter 1, and the carried message. The technical equipment needed for wave transmission and reception is broadly distinct from that used for modulation and demodulation. As a common rule, the carrier frequency should be much higher than the frequency of the signal to be carried, a ratio of 100 being typical.

The simplest modulation and the easiest to implement is binary: on-off, wave on–wave off. It is enough for telegraphy, using Morse's Code, for example. This was the first principle utilized by Marconi's Wireless Telegraph Company of the Italian engineer, after 1895. It was a very rough but very efficient modulation which is still used in some circumstances. However, the communication objectives became progressively more and more ambitious to transmit voice, sounds, images, data and led to design of ever new devices for more complex modulations.

2.1. Modulation and bandwidth

The simplest mathematical expression of a pure radio wave, a carrier, is the sine function. At a given place, the instantaneous electric field E can be represented by:

$$E = Eo \sin (\omega o\, t + \varphi o) \qquad\qquad [2.1]$$

where

– Eo is the field amplitude;

– ωo the angular frequency (which is equivalent to the carrier frequency fo using the formula $\omega o = 2\pi\, fo$);

– φo a constant depending on the place considered, called the initial phase.

The global expression in brackets $\varphi = (\omega o\, t + \varphi o)$ is the instantaneous phase and t is time, of course.

This basic expression does not give any information on the vectorial characteristics of E, such as polarization, but it does not matter here. It should be noted too that, for calculations, it is much easier to use formulae with complex quantities. However, this simple sine form is sufficient to introduce the practical notions that we have to deal with.

If the three parameters Eo, ωo and φo are fixed and stable, the wave is "pure" and does not carry any information. On the contrary, it is possible to make any of them vary according to the message. Thus, different modulations can be implemented:

– amplitude modulation;

– frequency or phase modulation, also called angular modulation.

During the first years of radio, until the 1970s, the modulations were mainly analog which means that for any message represented by a time varying signal v(t), such as the electric signal issued from a microphone, the chosen modulation parameter varies proportionally and instantaneously:

– Eo becomes Eo (1 + k v(t)) for amplitude modulation;

– φ becomes ωo (t + $\Delta\varphi$ v(t)) for angular modulation.

In these expressions k and $\Delta\varphi$ are constants which characterize the "depth" of the modulation, also called deviation. This modulation "depth" must not exceed certain limits as non-linear phenomena occur which distort the message.

After the 1970s, modulations preferably become digital. In this case the message should be coded at first. This is translated into digital frames made of binary numbers, or digits, which are formatted according to very strict rules. This translation, called information digitization, is the universal basis of modern information technologies. Now nearly everything is digitized: speech, sound, text, pictures, movies, data are processed in such a way to be transmitted, stored, modified, calculated, etc.

To carry a digital message on a radio wave, any digital value or binary word configuration should be converted into a definite radioelectric state of the carrier. In some other modern modulation schemes (such as COFDM, Coded Orthogonal Frequency Division Multiplexing), a definite combination of different carriers is associated with such a digital sequence. In any case the principle is the same: a definite bit sequence is associated with a definite electromagnetic state, which can be easily recognized without error and ambiguity, according to a strict logical algorithm.

As an example, let us suppose that we are able to produce at will a radio wave with four states of the initial phase $\varphi0$, $\varphi1$, $\varphi2$ and $\varphi3$. Let us decide that each phase state shall be associated with the different values of a two digit information block: 00, 01, 10 and 11, respectively. According to the digital sequence to be transmitted, a wave will be produced whose initial phase will rapidly change. Thus the digital sequence 010001101100… will be represented by a wave with a rapidly changing phase state: $\varphi1$ $\varphi0$ $\varphi1$ $\varphi2$ $\varphi3$ $\varphi0$…

A variant is to use a phase transition to signify the value of a digit rather than an instantaneous phase value. Thus the digital sequence will be represented by conventional phase jumps $\Delta\varphi 1\ \Delta\varphi 0\ \Delta\varphi 1\ \Delta\varphi 2\ \Delta\varphi 3\ \Delta\varphi 0...$ (QPSK (Quaternary Phase Shift Keying) Modulation).

Another common digital modulation is based on amplitude states. As an example a carrier can be modulated with 16 different amplitude/phase states (QAM (Quadrature Amplitude Modulation)).

A fundamental advantage of digital modulations, besides their capacity to be processed efficiently, is basically to ignore any transmission noise or distortion. If it is possible, on the receive side, to recognize without error the modulation states and to distinguish one from the other, the original binary message can be restored without any mistakes. Thus the received information is also strictly identical to the transmitted information. This is a unique advantage in comparison with analog modulations because these are progressively spoiled by transmission noise which adds to the original signal, after demodulation. This noise cannot be removed and the message degrades more and more after each transmission.

Here is not the place to study the different modulation properties in detail. A very large variety of techniques exists, in both analog and digital. However, a fundamental fact should be noted in any case which has been deeply analyzed by Shannon in his information theory, published in 1948. A "pure" wave, carrying no information, is seen in the spectrum as a single, infinitely narrow ray. On the contrary, a modulated wave spreads itself over a frequency band, a part of the spectrum, whose width depends on the modulation type and the information rate or band which is transmitted.

Let us consider the oldest and most simple modulation, amplitude modulation. A carrier wave whose frequency is F and which is amplitude modulated by a message whose maximum frequency is f, spreads over a 2f wideband centered on F. If the message is human speech, the useful message frequencies are approximately limited to a 4,000 Hz acoustic band. Thus the amplitude modulated wave occupies a 8,000 Hz wideband. If the carrier frequency is F = 1,520 kHz, the modulated wave spreads over the 1,516-1,524 kHz band.

This is the most straightforward case. Angular and digital modulations lead to more complex calculations for their spectrum width but in any case the necessary bandwidth is directly related to the information rate to be transmitted: the greater the rate, the broader the carrier's band.

For a particular modulation principle whose parameters are well defined and a referenced modulating signal, the bandwidth on which the carrier spreads itself can be calculated. This is why the radio spectrum can be formatted according to the different needs. When the modulation and information message are standardized, together they define a standard carrier bandwidth, a frequency slot, which can be used as a basic "channel" for this particular telecommunications service. Such basic channels are placed side by side in order to channelize the band dedicated to this service. Thus, in each channel, a carrier can transport a standard message.

Channelization is one of the most used tools by traditional radio regulation and spectrum management. It is now being challenged by some new broadband techniques which spread the spectrum on common widebands, but it remains, by far, the most popular management scheme. It is notably used to organize and share the spectrum over a definite geographical area, which is called "planning" (see Chapter 7, section 7.2).

As an example, the so-called Stockholm 1961 plan organized the 470-960 MHz band, dedicated to TV broadcasting in Europe. It shared this band between 61 usable channels, each one being 8 MHz wide and suitable for one analog TV channel. The adapted modulation principle is amplitude modulation with vestigial side band. This plan has just been updated (Geneva 2006) to cope with digital TV (DVB-T modulation standard), but the basic channel scheme stays as it was for analog TV. The Stockholm plan is familiar to anyone who has a TV set at home when choosing his TV program on a particular "channel".

In the same way, the frequency bands allocated to private mobile radio networks for taxis, fire brigades, maintenance companies, etc. are commonly "channelized" in 12.5 kHz wide frequency slots, each one being occupied by a phase modulated carrier. It should be noted, in such a case, that the corresponding service is often duplex (see section 2.4) and needs two channels for a single two-side communication. Thus two mirror bands should be planned with the same channeling scheme, being separated from one another by a "duplex gap", typically 10 MHz in Europe, in order that a conversation may use a bilateral up and down communication path.

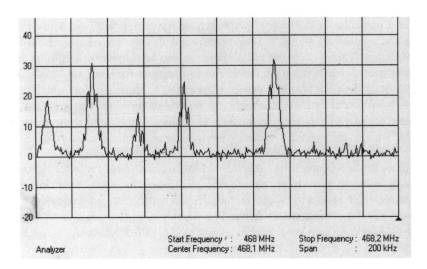

Figure 2.1. *Private radiotelephone channels. The band is divided into many individual channels for private radio networks, each channel being 12.5 kHz wide. Here five channels are active. The Y-axis scale is dBμV*

The international standardization bodies try to specify, for every new radio system, standardized messages and modulation schemes in order to define compatible basic communication channels which will be used as basic planning units by the Radio Regulations. The Atlantic City radio conference, in 1947, designated the different traditional modulations and elementary messages using symbols. As an example a double sideband amplitude modulated carrier for telephony is designated A3. Today these designations have a mainly historical value but they are still used in some traditional cases such as HF communications.

2.2. Bandwidth and noise

A receiving antenna catches simultaneously a great number of radio waves which utilize the same frequency band: the useful nominal carriers but also different signals coming from outside. These signals are produced by human activities or by natural radio sources. They enter the receiving station through the antenna main lobe or, more often, through secondary lobes and can be designated as interference noise. The electronic receiving chain, itself, generates a broadband noise, called thermal noise, which covers the whole spectrum and limits the receiver sensitivity. These different noises impair the useful signal at a variable level.

Thus it is necessary to select the frequency band which is admitted to enter the demodulator carefully in order that the whole useful signal be accepted but most of the other signals rejected, as far as possible. It is the function of dedicated electronic circuits, filters, to operate the selection between "good" and "bad" signals. A filter is characterized by its frequency transmission pattern which is suited to the useful spectrum and rejects unwanted signals on its sides.

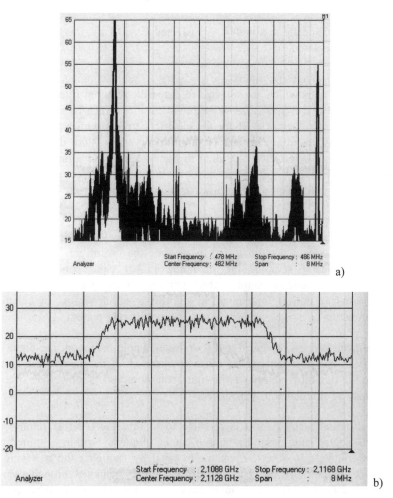

Figure 2.2. *Two types of modulations. a) Analog amplitude modulation by a TV signal. The channel is 8 MHz wide, centered at 482 MHz. b) Digital modulation by a UMTS signal. The channel is 5 MHz wide, centered at 2.1128 GHz. It can be easily seen that the analog modulation is much more complex and probably more sensitive to interferences. On the contrary, the digital modulation, using a spread spectrum technique, is very smooth. The Y-axis scale is dBµV*

From the very beginning of radio history, the need and importance of filters have been known. The most basic filters to be used are LC circuits made with a coil (L) and a variable capacitor (C), which are tuned to a frequency F:

$$F = 1 / 2\pi \, (LC)^{1/2} \qquad\qquad\qquad\qquad\qquad\qquad [2.2]$$

It is easy to calculate that with $L = 1\ \mu H$ and $C = 0.1$ nF, the tuning frequency $F = 16$ MHz is obtained.

Conventional electronic filters are designed using a complex combination of such elementary cells which are adjusted to shape the required filtering pattern. For microwaves, filters can be realized with tuned cavities which have, for waves, the same properties as electronic filters for electric currents. Other technologies are available such as electromechanical vibration of crystals. These different types of filters get different selectivity according to their technology: some are better than others. The "merit factor" of a filter made of interconnected electronic components is generally worse than that obtained from an integrated piezoelectric cell.

However, digital filtering using signal processing is becoming common. Thanks to the speed of new microprocessors, sophisticated real-time operations can be executed on the received radio signal to separate the useful part from noise. The receiver periodically samples the incoming global signal and measures the instantaneous amplitude of samples which is binary coded. These coded values are then processed by definite mathematical algorithms which realize, by calculation, the same function as a filter, according to a chosen filtering pattern.

For a given processing speed and a definite sample precision, the digital filter characteristics are fully determined by the parameters of the algorithmic software. Thanks to the flexibility and power of signal microprocessors, objectives other than filtering may be simultaneously achieved.

With different techniques it is possible to modify the receiver tuning, i.e. to select a particular band around a chosen center frequency within a limited tuning range. Of course, this objective can be reached by changing the receiving filter tuning itself. In the first radio receivers this was simply achieved through a variation of capacitance C in the front LC tuning circuit. It can also be obtained by modifying the dimensions of a microwave cavity. However, these techniques are too imprecise to obtain an appropriate selectivity. They are only used for a primary choice of the waves admitted in the receiver.

The most used selection technique is called "heterodyne" where the incoming radio signal is mixed with a frequency reference produced by a "local oscillator", in a "non-linear" electronic component such as a diode.

The local oscillator generates a pure frequency signal F1 which can easily be changed (it may be a fast electronically controlled frequency synthesizer). When a local signal crosses the non-linear component together with the incoming signal at a frequency F, a mixed signal is produced which is a copy of the incoming one but with the frequency translated around a center frequency Fi, called the intermediate frequency, with $Fi = F - F1$. With this technique, it is possible to use a high quality fixed filter, centered on the intermediate frequency and optimized for the useful channel standard. Thus the carrier (or channel) admitted into the demodulator is easily chosen by simply tuning the local oscillator frequency, F1. The filter tuning is never modified.

Let us consider, as an example, a good quality fixed filter centered at an intermediate frequency, $Fi = 70$ MHz. If a wave at a frequency $F = 410$ MHz should be demodulated, it could be mixed with the output signal from a local oscillator at $Fl = 340$ MHz. Mixing these waves will produce a copy of the incoming wave but centered at 70 MHz since $F - Fl = Fi$. It is easy to change the local oscillator frequency, let us say $Fl = 342$ MHz, to demodulate another wave around 412 MHz.

This technique is very efficient and flexible. It guarantees the same demodulation quality within the receiver tuning range. This range represents the spectrum domain where the carriers which the receiver can process are located. For a given tuning, the demodulator only deals with the narrow channel in the tuning range which is centered on the selected frequency and limited by the filter pattern around the intermediate frequency. It can be easily understood that such a technique is perfectly adapted to receive signals which are part of a plan with strictly formatted channels. The filter pattern is then suited to the channel spectrum profile.

Together with the useful signal, the filter allows all other signals whose frequency is included in its bandwidth to enter the receiver. All these signals are globally called electronic noise. It may be composed of various contributions, which can be precisely identified or not, depending on the receiving station environment. Often this noise does not show any particular structure, being only a broadband random signal. Such a noise is said to be "white" if its power density is uniform across the considered spectrum. This means that the white noise spectral power density being noted "n", the noise power in any band whose width is B is given by:

$$N = n \cdot B \qquad\qquad\qquad [2.3]$$

This formula is commonly used to calculate the noise power which enters a demodulator after the receive filter whose bandwidth is B. It is the most frequent case when no specific interference occurs in the receiver vicinity. Any receiving system made of an antenna and an electronic receive chain can be characterized by a typical quantity called "noise temperature", T, corresponding to the white noise spectral density which it generates by itself.

$N = k \cdot T$ with k being the Boltzman's constant: $k = 1.38 \cdot (10)^{-23}$ joule/kelvin   [2.4]

In such a white noise, it may be mixed with the thermal noise generated by the electronic receiving chain, the environmental noise captured by the antenna through its primary and secondary lobes. The noise temperature T can be broadly considered as constant and stable for a particular receiver type on a given area where no specific interference is met: this determines the receiver sensitivity.

Let us consider, as an example, a satellite system receiver with a noise temperature of 200 kelvins. A 25 MHz wide filter will admit a noise power N:

$$N = 1.38 \cdot (10)^{-23} \cdot 200 \cdot 25 \cdot (10) \exp 6 \text{ watt,}$$
$$\text{which is } N = 6.9 \cdot (10)\exp\text{-}14 \text{ watt, or } 0.069 \text{ picowatt}$$

It will be seen, in Chapter 10, that the idea of giving more freedom to use radio devices which create a broadband spread spectrum has recently appeared. In their vicinity, the radio signal looks like white noise. Thus the general environment could be characterized anywhere by an "interference noise temperature" which could be limited by regulation. The noise temperature which would be met anywhere would then be the sum of the thermal noise temperature and the interference noise temperature.

In a number of circumstances, however, the most important noise is produced by one or several identified interferences, typically due to near-by transmitters. In such a case the noise can be calculated by applying to the interfering signal the same rules and equations as the nominal useful carrier (see section 2.5). This situation is of primary importance for radio regulation procedures as seen later.

2.3. C/N (or C/I) and S/Nm

This section's strange title introduces a first summary of all notions previously described about the relationships between useful signals and noise.

Any communication is impaired by noise. The transmitted message is never the original one. Let us try to clarify how such degradations appear.

We have already discussed the basic difference between the radio "transmission", which concerns the carrier wave, and the radio "telecommunications" which concerns the message carried by a modulated wave. Should we propose here a book about telecommunications, we would explain the different possible interactions between these two aspects and we would show how some propagation distortions and noises may affect the carried message. However, concerning only spectrum management, it is possible and realistic not to concern ourselves with these phenomena. This is fully justified, notably, for digital communications where the two issues, carrier transmission and telecommunications, are nearly independent in normal conditions.

Thus, we can separate two segments in the global radiocommunication link:

– the transmission segment which concerns the carrier wave;

– the message processing segment.

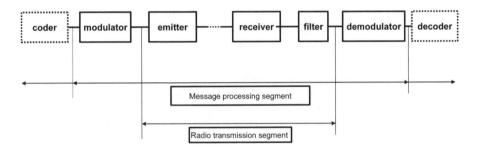

Figure 2.3. *Radiocommunication link. In a radiocommunication link two segments should be separated: the radio transmission segment which concerns the carrier wave and the radio noise. This is characterized by the C/N or C/I ratio at the output of the receiving filter; the message processing segment which includes the information coder and decoder together with the carrier modulator and demodulator. Its quality is described by the S/Nm factor*

The transmission segment extends from the modulator output to the demodulator input, the receive filter being included.

The message processing segment includes the modulator-demodulator couple with, possibly, another coder-decoder couple to shape the information signal in a form adapted to the transmission segment.

The transmission segment crucial point is the demodulator input, after filtering. At this point, the incoming useful carrier appears with a superimposed radioelectric noise. The carrier power level, designated here as C, is directly related to the received radio power (Pr as seen in Chapter 1). The noise power level which passed through the receiving filter is designated as N. It can be preferred to designate it as I (for interference) if the incoming noise is mainly due to an identified external signal whose power exceeds the thermal noise (this situation is becoming more and more common with the growing sharing and re-use of radio frequencies to improve the spectrum efficiency).

In any case, the ratio C/N or C/I is fundamental to assessing whether a radio transmission is possible or not. In most modulation-demodulation techniques, there is a limit value, a threshold, which separates an effective transmission from an ineffective one. As examples, for a digital radio modulation, this limit is met when the different carrier radioelectric states cannot be clearly separated without any error. For frequency modulation, this threshold corresponds to a value under which the radio noise sharply degrades the message quality.

It can be easily seen that the useful signal power C should be significantly stronger than the noise N to be demodulated. As orders of magnitude, if their ratio is greater than 20 dB (a power ratio of 100, an amplitude ratio of 10), it can be thought that the noise will not disturb the demodulation. If their ratio is about 10 dB (an amplitude ratio of 3.2), some problems will surely be experienced. If the noise power is higher than the useful signal, to a point where the useful signal is completely "drowned", the job may be thought impossible. In fact, in any circumstances, a detailed technical analysis is necessary to reach a conclusion. Even in this last worst case, some techniques using signal processing can "extract" a useful signal from a great amount of noise. Moreover, this is the very principle of modern transmission systems using spread spectrum wideband modulations.

It should be noted that noise is often a random radio signal. Its instantaneous amplitude varies around a value called "rms (root-mean-square) amplitude", proportional to the noise power square root. This instantaneous amplitude may be at times much higher than the rms value. It is this instantaneous amplitude of noise which should be considered for the demodulation of narrow band systems since it may exceed the carrier's, even during a small percentage of time.

Let us suppose that the C/N (or C/I) ratio is high enough for a fair demodulation. We may be interested by the quality of the demodulated message S.

In analog communications, the demodulated signal S appears as affected by a noise Nm (Nm, "demodulated noise", is definitely different from N, "transmission noise"). The S/Nm ratio, after the demodulator, between the message power S and the demodulated noise power Nm, measures the telecommunications quality. It can be calculated from C/N by some equations which depend on the modulation principle.

For digital communications, we have already explained that when the demodulator performs above its noise threshold, it reproduces the original message identically, without errors. Thus, it may be considered that, in such conditions, the binary message does not suffer any distortion by the telecommunications.

This is true; however, the digital representation of any original analog signal cannot be perfect. The usage of coder-decoders which realize the translation may somewhat degrade the original message. A musical melody binary representation may be as accurate as necessary if the binary stream is fast enough. On the contrary, if the binary speed is limited, a specific distortion is created, called quantization or coding noise, which is due to the approximation between the original message and its coded version. Thus, the telecommunications link is still characterized by an S/Nm ratio but in this case Nm has no direct relation with the transmission link (in fact such a relation indirectly exists, since when the bit stream is made greater for a better message fidelity, the necessary bandwidth must be proportionally increased and the transmission noise is increased too).

We can consider that, in the future, when all communications are digital, the spectrum management will only be concerned by the C/N or C/I ratio which affects carriers and will ignore the coding-decoding process. Nowadays, however, many analog links still exist and both C/N and S/Nm ratios should be assessed to evaluate the link quality.

For a number of radio communications, primarily directed to customers, such as mobile or broadcasting services, using omnidirectional or roughly directional antennae, it is not possible to precisely evaluate every individual radio connection. The global communication quality can be statistically assessed and the individual link quality may be different from one place to another. It can be considered that with standard receivers, in conformance with state-of-the-art technology, in a radio environment characterized by a definite thermal noise without particular local interference, a good service quality will be achieved if a minimum radio field is created on the concerned area.

In such cases, interferences are often generated by the network itself, mainly due to multipath propagation, co-channel interference, reuse of frequencies in a cellular scheme and so on. They have to be modeled at the very beginning of the network design and planning. It can be considered that when they stay lower than a definite value on the service area, they do not impair the service quality. The same attitude should prevail for other eventual interferences.

On the contrary, for radio fixed services or satellite links, the calculations may be much more accurate and take into account a number of individual parameters relevant to each situation. Each link can be modeled and assessed using the full precision of the power equations seen in Chapter 1.

$$C = Pe.Ge.Gr/Ap$$

$N = k.T.B$ for thermal noise and I evaluated from each source of interference, if existing.

Let us consider the basic quality of any link, when no interference occurs:

$$C/N = Pe \, . \, Ge \, . \, Gr \, / \, Ap. \, k \, . \, T \, . \, B \qquad\qquad [2.5]$$

It can also be written:

$$C/N = (Pe.Ge) \, . \, (Gr/T) \, / \, Ap \, . \, k \, . \, B$$

Two interesting expressions can be pointed out.

[Pe.Ge] is the equivalent isotropic radiated power (EIRP), already seen in Chapter 1. It only depends on the transmitting station. It is possible to balance the two Ge and Pe parameters to get the same EIRP objective.

[Gr/T] is the "quality factor" of the receiving station, antenna and electronic receiver together. It is an important parameter to globally evaluate if a receiving station technology is appropriate. It depends on the antenna size and the receiver sensitivity. Here too, a balance can be made between Gr and T for a given [Gr/T] objective.

These quantities, which should preferably be expressed in logarithmic units, are much used for regulation and standardization, being more global and flexible than Pe, Ge, Gr and T. The possible trade-off between the technical choices which they offer helps to find the best economic compromise and also to facilitate the electromagnetic compatibility between systems using the same frequency band.

As an example, the first great INTELSAT antennae (Class A, C band) which were the reference for commercial satellite communication stations during the existence of this international organization were about 30 meters in dish diameter. If such stations were built according to the state-of-the-art technologies, they achieved a quality factor $[Gr/T] = 40.7 + 20 \log (F/4)$ dB/K, with F expressed as gigahertz. This quality factor can be obtained with a 57 dB gain and a noise temperature of 43 K, but it may also be achieved with a 61 dB gain and a noise temperature of 108 K. The impact of such choice on the receiver technology was a major one.

Another interesting expression is obtained when retaining only in C/N the factors which are independent of the modulation, and neglecting the k constant. There remains:

$$[C/T] = Pe. \ Ge \ . \ Gr/Ap \ . \ T = [EIRP] \ . \ [Gr/T]/Ap \qquad\qquad [2.6]$$

This quantity [C/T] is useful because it can be cumulated from path to path in a radio system with intermediate relays, without reference to a particular carried message, if no demodulation occurs from end to end. This is, for example, the case for satellite communications with up and down paths between the Earth and the satellite, or for microwave links which can cover hundreds of kilometers with regularly spaced relays. This cumulative property gives an idea of the transmission path degradation after several relays.

$$1/[C/T] = \Sigma i \ \ 1/[C/T]i \qquad\qquad [2.7]$$

The inverse of the global [C/T] is the sum of the inverses of elementary [C/T]s.

This quantity [C/T] is also broadly used for radio coordination calculations (see Chapter 7, section 7.3), notably for satellite systems, when it is necessary, for example, to assess the degradation which an existing system may suffer from a new incoming one in its vicinity. This degradation is evaluated as a noise temperature increase ΔT and the radio regulation imposes that the relative degradation $\Delta T/T$ be limited to a small percentage.

2.4. Multiplex, multiple access

A radio network frequently uses a same spectrum resource to carry a great number of independent messages simultaneously (or apparently simultaneously). This technique is called multiplexing and many different types of multiplex are implemented. Historically speaking, a multiplex is a fixed collection of independent messages which is point-to-point transmitted between two stations, as a whole. Variants of multiplexing are multiple-access techniques where the different messages carried by a common radio resource are exchanged between a number of different stations in a service area. However, these techniques are more and more mixed and may be considered as common in a digital environment.

The most basic multiplexing (or multiple access technique) is used for alternate speech communications. For a number of simple usages, dialogs must be established between two or more parties. Typically, one speaks while others listen. Later an answer comes from a partner. This type of communication is compatible with a unique radio channel which can be successively used by the different parties. Such a channel is called a "simplex" channel and its alternate use is manually controlled by participants. The popular walkie-talkies are based on such a principle.

Simplex is of course not very appropriate for a comfortable dialog between two parties. In this case it is preferable to use two channels, one for each direction, creating a "duplex" link. A duplex is well adapted to permanent bilateral communication. It may be noted that simplex, although an imprecise procedure, is more spectrum efficient than duplex.

Many small private networks use 6.25 or 12.5 kHz wide phase modulated channels, typically in bands around 150 MHz or 400 MHz, either as simplex channels or as duplex ones. In this case the paired duplex bands are preferably 10 MHz apart in Europe. As an example the band 446.1-446.2 MHz is harmonized for short distance simplex digital walkie-talkie equipment.

Let us now consider a permanent point to point link set between two fixed stations where many different messages are transmitted. It is appropriate to combine these elementary messages into a unique global message, a multiplex, which can be carried as a whole, being considered as a single message from a radio point of view.

The analog radio links, in the past, carried telephony multiplexes made of individual telephone channels, frequency translated and placed side-by-side in a broad frequency "baseband". Each channel occupied a 4 kHz slot. A 60 channel multiplex could then be considered as a unique message placed in the 12-252 kHz frequency "baseband". The radio carrier was modulated by such a global message and was unaware about the individual telephony channels inside.

This analog technique was replaced during the 1970s by digital multiplexing. Each telephone message being coded at a speed of 64 kbit/s, a 30 channel multiplex with two auxiliary service channels was organized as a digital frame with a 2.048 Mbit/s binary rate.

Now, with more sophisticated message coding techniques and the modern dominance of packet routing of information (IP protocol), the multiplex content management becomes dynamic and statistically controlled. The multiplex data stream is used to carry information blocks from the different messages, depending on their instantaneous needs and priorities. Then the content of the multiplex, as a transport digital frame, varies in real time and can be fluently managed with the best efficiency.

The point to point multiplex (now becoming transport digital frames) can be standardized, looking like ship containers, with hierarchical capacities. They may be carried by different telecommunications vectors: fixed terrestrial radio links, long distance cables, satellite, etc. For this purpose some radio bands are formatted or planned according to the size of standardized radio channels, each one capable of transporting one such multiplex.

More and more radio networks are distributed, which means that they connect several stations located in a service area. Some are said to be point-multipoint when a fixed master station, called a base station, is connected to a number of secondary stations or slave-stations. This is typically the case of mobile public radio telephone networks or satellite networks. However, pure multipoint networks also exist where any station may directly communicate with any other one, being all equal. The methods which share a common radio resource between the different stations in such distributed networks are called multiple-access.

Three basic methods are commonly used:

– frequency division multiple access (FDMA);

– time division multiple access (TDMA);

– code division multiple access (CDMA).

Other methods can also be used, such as space division multiple access, and several methods can be combined in a particular network.

However, opposite these methods which control all communications with a deterministic procedure, one of the multiplexing protocols which found favor at the beginning of radio data networks was ALOHA (whose principle was reused in the ETHERNET protocol for cable networks). With ALOHA, all transmitters can freely transmit on the same frequency, at any time. Due to this simple strategy, some "collisions" may happen when two stations send their carrier at the same instant but they can be accepted if their occurrence probability is small, which is the case when the global traffic load is low. These collisions, when they happen, are detected and corrected by an automatic repetition of the jammed data.

Some modern radio local area networks (RLANs) make use of such collision avoidance techniques. As an example, IEEE 802.11 is a family of standards designed by the American association IEEE, dedicated to RLANs which connect various equipment to the Internet network. The most popular is 802.11b (WiFi), oriented to personal computer configurations, for a maximum data speed of 11

Mbit/s (about 3 Mbit/s in practice). The frequency band is 2,400-2,483.5 MHz, divided into channels which are not equally accessible in all countries. The modulation technique is a digital phase modulation (PSK) and the spectrum is spread by different techniques such as frequency hopping (FHSS) or direct sequence (DSSS). The multiple access of different stations to a common radio resource is controlled by a distributed ETHERNET type protocol to avoid collisions.

The inventiveness of engineers never stops. Let us come back to the most common procedures on multiple access networks.

Frequency division multiple access divides the available radio frequency band into many sub-channel units which are temporarily allocated to the elementary communications, according to the needs at each moment. When established, a sub-channel communication is considered as a stable point to point link: the only difference is that every link is a temporary one. The global radio activity is statistically described in tight relation with the communication traffic.

Such a procedure which allocates a limited radio resource only to an active communication is a common feature in the multiple access networks because it drastically improves the spectrum efficiency. As a drawback, it makes radio management, which becomes real time, more difficult. However, computers help…

In time division multiple access, each active station transmits alternatively the same carrier during a short time span. Practically, only digital messages can be treated by such a method. If the mean information rate to be transmitted by one station is N bits/s, T is the cycle period when each station is authorized to transmit, τ is the time allocated to this station during every cycle, the instantaneous binary rate of a transmitted digital frame should be at least $N.T/\tau$.

In such a multiple access method, the receiver has to cope with wave bursts coming from the different transmitting stations. All of them, at the same frequency, are time distinct, each one carrying a block of binary information. Thanks to temporary memory stores, the transmitting station cuts the original message into formatted digital packets and the receiver reassembles them to restore the message. Complex real-time logic protocols are associated with these operations and the corresponding data should be carried along with the message on each link.

Code division multiple access is the most recent technique. It is based on the orthogonality properties of some binary code families. Orthogonality means that it is possible to clearly recognize and recover two different codes in the same family, without error, and that digital receivers can extract from a noise context a signal which has been coded according to a particular scheme. In a CDMA network, every station transmits simultaneously on the same carrier frequency but each one using a different code from the same orthogonal family. All these emissions create a kind of radio noise which should be interpreted by every receiver. To demodulate a particular message, the receiver should process this "noise" with a digital signal processor. It has been provided with a "key" which is the particular code associated with this message.

In this context too, the instantaneous digital rate of each transmitter is much higher than the mean information rate to be carried. The coding process multiplies the information rate as a function of the number of codes which are available in the family. Then the carrier spectrum is "spread" over a broad frequency band.

All these techniques are more or less sophisticated multiplexing where a common radio resource is shared by different independent messages. It must be understood that in all cases, as for point to point links, the necessary frequency band to establish the radio network is directly related to the global information rate which is transmitted inside the network between the participant stations. Let us give some figures as examples chosen from different recent European radio mobile telephone and digital TV standards.

FDMA-TDMA: the GSM case

GSM, the first digital mobile telephone standard in the world, uses several multiple-access and multiplexing techniques simultaneously.

The basic radio channel is 200 kHz wide. Then the GSM allocated primary bands, 890-915 MHz (for mobile to base station links) and 935-960 MHz (for station to mobile links) is divided into 125 elementary channels (124, in fact, due to protection guard bands).

A radio carrier propagated from a base station to the mobiles running in its vicinity brings digital frames, each one made of 8 elementary time slots whose duration is 0.577 ms: then a frame lasts 4.615 ms.

Each time slot counts 156.25 bit times which are notably used for 116 information bits and 8.25 bit guard times to separate the slots. Then the useful digital rate allocated to each communication is equivalent to 116 bits, every 4.615 ms, which is approximately 25 kbit/s.

A carrier capacity is approximately 200 kbit/s (8 * 25 kbit/s)

In the reverse direction, from mobiles to a base station, each mobile transmits a burst signal whose duration is 0.577 ms, every 4.615 ms.

According to this scheme, the basic spectrum management structure for GSM is FDMA since the spectrum is divided into independent 200 kHz channels. However, these basic channels are used by TDMA multiplex and multiple access techniques since eight communications are alternatively using a common channel.

CDMA: the UMTS case

UMTS, a third generation mobile radio network, uses a CDMA procedure. In its W-CDMA version (W stands for wideband), the basic radio channel is 5 MHz wide. Each useful message whose rate is d bit/s is combined (by a binary multiplication) with a pseudo-random code taken from an orthogonal family, which is strictly associated with this message to achieve a global transmission rate D bit/s. The D/d ratio is the spreading factor since this operation spreads the modulated wave spectrum to occupy the full radio channel bandwidth. Practically, the useful information rate which is offered to a communication may be as high as 384 kbit/s and the transmission rate is 4.096 Mbit/s. Up to 512 codes are available for the same number of communications.

As already explained, according to CDMA principles, different carriers using the same channel are simultaneously transmitted. Their spectrums are added together and produce a white-noise-like radio signal (see Figure 2.2). To demodulate a particular message, the receiver should generate the corresponding pseudo-random code and use it as a key to extract the good information from this noise, through a "deconvolution" process.

Digital TV multiplex in Europe: DVB and MPEG standards

DVB (digital video broadcasting) is a family of broadcasting standards for digital TV, adapted to the different transmission media: satellite, cable, terrestrial radio, etc. DVB-T is dedicated to terrestrial broadcasting and can replace a standard analog program in a 8 MHz channel of the European TV plan.

The carrier modulation is COFDM (coded orthogonal frequency division multiplex) which is well adapted to difficult propagation conditions with multiple paths and echoes. The 8 MHz spectrum resource is divided into a large number of elementary carriers (6,817 in the 8 k mode, 1,705 in the 2 k mode) which carry a digital modulation (QPSK or QAM). The carrier scheme is made of successive elementary frequency/time cells where a particular combination of elementary carriers is activated. Two cells are separated by a guard time. The elementary carrier configurations are orthogonal from one cell to the following one.

On such a digital channel, several TV programs are statistically multiplexed. They are coded according to different standards, more or less efficient, for various picture qualities. The basic choices are MPEG2 and MPEG4 (MPEG as Moving

Picture Expert Group), which convert a TV picture into a digital stream with a data rate between 1 and 5 Mbit/s.

Chapter 10 provides a more detailed scope of some modern digital modulations which may lead to important changes in spectrum management practice.

2.5. A balance between carrier power, noise and interferences

It may be useful to come back, with more details, to the matters of "noise" and "interferences" which are probably the most critical notions for spectrum management, and their relations with the carrier power.

As mentioned in the introduction, the main objective is to reuse as much as possible the radio spectrum from place to place, from one service to another, to get the best efficiency.

It has been seen, in section 2.3, that radio communication is possible and only possible if the C/N ratio between the received carrier power, C, and the noise power, N (including, if any, interferences), as admitted in the receiving filter, in front of the demodulator, is high enough. Using an analog modulation, the quality of the received message depends directly on this ratio. Using a digital modulation, this ratio controls the digital error rate which is the key factor for the communication capability, i.e. above a certain threshold, the communication is perfect, under, it is impossible.

Thus, nothing is said about the absolute value of C, or N. Only their ratio matters: a balance has to be made between C and N for every radio system, anywhere, to find the optimized parameters for global satisfaction.

Two opposite principles should be kept in mind:

– to work, a new radio system should cope with the existing radio waves in its environment which it perceives as a noise;

– any new radio system must not generate radio signals which would harm existing systems using the same frequencies, by interferences.

It is easy to understand the necessary equilibrium.

Let us suppose that a new system is introduced but with a noise N, as observed at the receiver input, being estimated too strong for a good communication. It can be thought that a solution is to increase C. This will probably solve the problems of this system.

However, the new carrier is seen as a noise by other existing systems: if its power is incautiously increased, it will produce harmful interferences.

Let us add that it is a bad policy to ever increase, without any control, the mean and general radio field existing on the Earth.

The different parties have to find together the right parameters which satisfy existing and new systems. This can be obtained by editing some standards, general rules and plans which must be followed by all systems (general electromagnetic compatibility approach). This can also be obtained, on a case-by-case approach, by detailed individual studies (coordination procedures).

Thus, all spectrum management bases are summed up.

C is always calculated from the transmission equations as seen in Chapter 1, notably [1.9], taking into account statistical values of the path attenuation Ap.

N is often more difficult to evaluate since it may include many components from many sources. However, noise prediction and modeling are fundamental to designing a new radio system.

At first, a background noise exists. This natural "thermal noise" is spread all over the radio spectrum with a uniform power density in any particular band, being called "white noise". It may be captured by the receiving antenna and is also generated within the receiver electronic equipment itself. It can be predicted, modeled and calculated from the beginning as a function of the technical characteristics of the receiving station.

Noise is also produced by all radio transmitters which are at work in the same band, in the vicinity of the concerned receive station. Generally, this noise can be statistically estimated. It can also be predicted and managed in the context of networks whose frequencies are globally engineered. This is the case of CDMA networks, in principle. It is also typical of what we have called "network noise" when planned channels are used on an area with engineering tolerances and margins which accept some limited and controlled interferences from one station to another. Cellular networks and broadcasting planned networks are commonly built on such a basis.

These different noise sources are broadly managed from the beginning as true parts of the equipment design and network engineering.

However, particular stations may produce specific interferences which have to be studied case-by-case from their individual parameters and characteristics. This is the main objective of the coordination procedures.

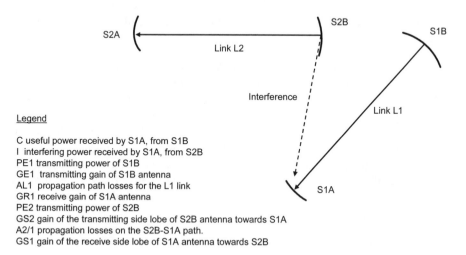

Figure 2.4. *Electromagnetic compatibility of two radio links L1 and L2 using the same frequency. Interference from L2 on L1*

To calculate the radio compatibility between two links using the same frequency, the useful signal level coming from the corresponding transmitting station, C, and the interfering signal coming from the other transmitting station, I, should be estimated for each receiving station. This interference is commonly radiated and caught through the antenna secondary lobes.

Let us study two radio systems L1 and L2, using the same frequency, as an example two stations of the fixed service working on a same channel: they may interfere. To assess their compatibility, we should calculate at the receiving station S1A the received power of the useful signal coming from S1B, which is the nominal transmitting station:

$$C = PE1 . GE1 . GR1 / AL1$$

Then the interfering signal coming from S2B is evaluated:

$$I = PE2. GS2 . GS1 / A2/1$$

The value of the C/I ratio when compared with the C/N ratio which applies to the nominal link determines whether the disturbance caused by the L2 link to the L1 link is negligible, acceptable or harmful. It will prove the electromagnetic compatibility between the two links, or not. Such calculations are the basis of coordination procedures. Of course the same calculation should be done to evaluate the disturbance of L2 due to L1.

When a radio system is installed and works properly, it may at some time be disturbed by unexpected external radio sources which generate carriers at a level higher than foreseen. These interferences can be estimated unacceptable when they deeply degrade the link quality or even interrupt the communication. If these interferences are very brief accidents, they may be accepted because no radio system can be 100% time proofed. However, if they are permanent or frequent, the capacity and availability of the disturbed system may be threatened and it is then appropriate to make use of the radio regulation which protects the rights of the parties through the assignment and coordination procedures. This will be introduced and discussed later.

Many reasons may cause unexpected interferences.

Some are natural and independent of any human activity. As an example, some propagation conditions can considerably lower the protection normally provided from a distant transmitting station by path attenuation. For satellite communications, a satellite-earth station link can be periodically interrupted when the sun, which is a strong radio source, crosses the receiving antenna main lobe.

Other interferences may result from technical failures or definite regulation offences. If any important transmission parameter is faulty, such as frequency, transmitting power or frequency deviation, there is a good chance that another user will face harmful interferences. However, such situations are clear and can be easily corrected.

It is more difficult to detect and cure some interferences due to accessory equipment. A typical situation results from bad engineering of radio sites where spurious signals are generated either by transmitter saturation or intermodulation. As an example, let us consider a site where there are several powerful radio transmitters, using frequencies F1, F2, F3, etc. Any non-linear device situated in the vicinity, such as a rusted metal piece, may create spurious radio signals, on frequencies $2F_i - F_j$ or $F_i + F_j - F_k$ which are in the same band as the nominal carriers.

Suppose three carriers, F1 = 400 MHz, F2 = 402 MHz, F3 = 404 MHz, are transmitted from the same side. They may locally combine and create an erratic signal at 398 MHz (2F1-F2) or 402 MHz (F1 + F3 − F2).

When interferences are found unacceptable, the radio regulation gives the opportunity to initiate a formal administrative procedure through a claim for interference. This claim will be studied from a technical point of view to determine the reasons for interference and from an administrative point of view to determine the responsibilities of the parties, according to their rights and duties. In every country, dedicated bodies have a mandate to deal with such procedures.

Chapter 3

Geography and Radio Communications: Radio Network Engineering

Radio communications are necessarily implemented on the Earth taking into account a number of geographical constraints: physical, political and human. It is thus useful to pay some attention to these questions which are the basis for regulation of electromagnetic laws and information theory.

The relief, land and sea, continents and oceans, borders, town and country, atmosphere and space are considered as a background when the rules for spectrum resource management are written. They directly influence the legal, administrative and technical dispositions and procedures which are enforced. They are among the main parameters to be evaluated when designing a new radiocommunication application.

Thus, considering the service to be offered to users over a definite area, the available frequency bands, the technical and regular constraints, investment costs and customer market, the operator must adopt an appropriate strategy to implement the radio network in the field. This strategy will often be guided by geography. The regulator who may supervise the operator activities and the network evolutions has also to care about this in its decisions, when issuing licenses and sharing the spectrum resource to insure a good coverage and fair competition.

3.1. Regions and countries

In order to facilitate international spectrum management and introduce some flexibility, the ITU (International Telecommunication Union), which is a United Nations organization, has divided the world into three Regions (with R as a capital letter to mark a difference from the usual meaning). They are only administrative and technical areas for frequency management purposes.

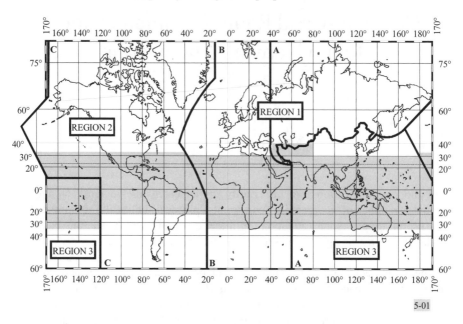

5-01

Figure 3.1. *The Regions in the Radio Regulations. Region 1 covers Europe, Africa, the Middle East, Russian territories in Asia and republics bordering these countries. Region 2 covers the Americas. Region 3 covers Asia and Oceania*

However, the exact division between the Regions may be locally complicated by political considerations which are not considered here.

This world division takes into account geopolitical constraints but also fits with the natural structures of continents which can be considered as independent, from a radioelectrical point of view, for waves whose frequency is greater than 40 MHz (the approximate limit for ionosphere reflection). With a frequency lower than this value, due to this reflection phenomenon, waves can be propagated worldwide, which makes the Regional concept inappropriate.

On the other hand, for frequencies higher than this value, the propagation distance remains limited: oceans and deserts isolate the Regions with such a loss margin that these areas can be considered as independent and managed accordingly.

Consequently the Radio Regulations, which set the rules for spectrum management, distinguish in the world frequency allocation tables the three Regions, placed side by side on three columns. However, for the greater part of the decametric band (from 3 to 40 MHz, roughly speaking), considering the ionosphere phenomena and a worldwide harmonization of applications, the international regulation does not distinguish between Regions.

Inside a Region and except for some local arrangements which are mentioned in footnotes, all countries comply with the same frequency allocation basis for the different radio services. This does not mean that every country in the Region uses the allocated bands identically: the regulation gives the possibility of making different choices between some services, when they are compatible. Moreover, a number of different actual applications are relevant to a single regular service. In addition, national or regional arrangements can be locally made if they do not impair the rights of any other country, as set by the regulation (see Chapter 4).

In fact, a fourth Region does exist: space, which is considered a common property of mankind. The frequency bands dedicated to space radio services are also allocated in conformance with the three Regions. However, their actual management is made according to common rules and procedures, broadly independent from the Regions. This is one reason why the highest frequency bands which are more easily used by space services than terrestrial ones, are mostly allocated worldwide. Today, nearly all bands higher than 30 GHz are allocated with no difference between Regions.

To sum up, the Regional structure is only pertinent for radio bands between 30 MHz and 30 GHz which are, however, at the very heart of the useful spectrum for most applications, notably for terrestrial ordinary communication services. It should also be stressed that countries in different Regions are not in favor of different allocations at any price. On the contrary, a common objective, in ITU, is to achieve an international harmonization for a better development of radio services whose market is growing fast on a worldwide scale. From this point of view, market liberalization, the growing influence of mobile services, the increase of long distance travel for leisure or business are powerful incentives to an ever growing harmonization of spectrum use all around the world. A number of bands are already allocated worldwide with no Regional exception.

Having in mind these constraints and long-term trends, also knowing its margins for freedom, each country may design and issue its own frequency allocation table, where every band is allocated to one or several particular services with dedicated bodies to control this utilization, such as a ministry, governmental agency, operator regulator, etc. Within Regions, this secondary allocation may be organized and harmonized through voluntary cooperation. In Region 1, as an example, the national allocation tables are harmonized through CEPT decisions and recommendations (see Chapter 7).

Thus, three geographical levels should be distinguished for general spectrum management.

The world level is the ITU's domain which has divided it into three Regions which are technical and administrative entities, usable for radio regulation but without any other strong political meaning. Spectrum regulation at world level is expressed in a document called the Radio Regulations which is elaborated by periodical World Radio Conferences (WRC). It is considered an international treaty.

When a question is relevant from ITU but concerns only a limited geographical area, a Regional conference may be organized where all countries of this Region are invited. The conclusions issued from the meeting are included in the Radio Regulations but are only a rule for that Region.

An "informal" regional cooperation may be organized and managed by different countries having common views on a spectrum strategy. They decide to cooperate on a voluntary basis and may have the ambition of a common harmonized radio domain for some services which could extend to a large area such as a continent or a broad geopolitical entity. Such a structure has no direct and legal connection with the ITU Regions but, of course, can play an important role in a world or regional conference if all the concerned countries express the same views and back them. This is typically the purpose of the European Conference for Post and Telecommunications (CEPT) which harmonizes the spectrum for most countries in Europe and some surrounding countries, although it has no effective regulatory powers.

The basic level is national. Every nation is in charge of managing frequencies in its territory. It does so according to its political and economical goals but with respect to international decisions. States are legally responsible for the use of the spectrum within their borders and the interferences which they create outside, if any. Thus, national borders are considered as a basic fact in the Radio Regulations, especially for frequency assignment and protection against interferences. Legally, nations are the basic actors of spectrum regulation, even with many constraints from abroad, and make up the ITU Members.

Some management levels may exist at a lower level than the nation, typically in countries with a federal structure such as the USA or Germany. Their constitutional law may share the spectrum regulation competences between a federal level and local governments for some customer radio services such as broadcasting. However, these national arrangements are not considered at an international level.

Some countries may also combine some of their competences in a political union and radio communications may be included in these Union competences, at least for some services. This is the case for the European Union which stated a decision in 2002 giving competences to the Union bodies for particular frequency matters. The relevant decisions are then mandatory for the participating countries. However, up to now, such unions are not considered formal Members at ITU level.

3.2. Radio implementation in the field

In normal conditions, the physical laws which rule radio waves make them propagate along straight lines. This is the reason why terrestrial radio communication networks, which link stations on the Earth's surface without any intermediate space, should be designed and built in accordance with the actual geographical features of the country. They have to cope with the Earth's rotundity which objects to direct links over some distance and also with the relief which is fundamental when choosing sites where radio stations will be built.

Some other geographical features have also to be considered such as seas and deserts where propagation conditions may be unusual: they may affect the protection against interferences.

For terrestrial communications, men have mostly used natural or man-made high sites. Since Antiquity, optical communication paths have been established between such high positions which are mutually placed in a direct line of sight, without intermediate obstacles, not too far from one another, looking for the best compromise between range and reliability of vision.

Radio links have the same objectives as optical links and frequently use the same recipes, to a point where radio stations occupy the same places as their optical ancestors. To increase their propagation range, to keep clear of obstacles, it is easier to install the radio stations on high sites and this is the most natural choice for engineers. However, modern radio networks do not take this as a rule. For high density radio networks, as an example, transmission sites which are not too high may be chosen in order to reuse frequencies more easily. In addition, for protection objectives against interferences, remote places surrounded by natural obstacles may sometimes be preferred, typically for signal reception from space.

The Earth's rotundity, by itself, is a major obstacle to direct line-of-sight communications. Let h be the observer's altitude above the ground and R the Earth's radius (6,350 km), thus the horizon distance L which is the limit for the direct sight range to a receiver on the ground is

$$L = (2 \text{ h.R})^{\frac{1}{2}}$$

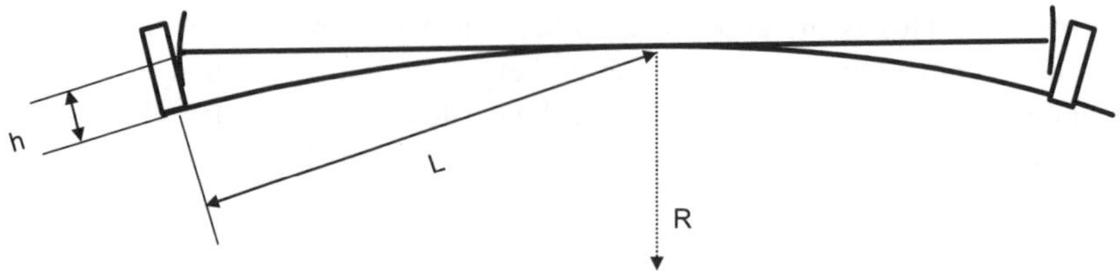

Figure 3.2. *Direct sight range and Earth's rotundity. Note: heights are greatly enlarged*

From a 10 meter high tower, the sight range is 11 km. This figure gives a rough idea, not taking into account any propagation losses or refraction phenomena, of the maximum range of a radio transmitter placed at such a height, in the 40 MHz to 30 GHz bands.

A 300 meter high tower, such as the Eiffel Tower in Paris or the big TV towers built in some cities, covers a circular area whose radius is about 60 km. The same transmitter at the top of a 2,000 meter high mountain reaches receiving stations up to a distance of 150 km.

However, in fact, due to atmospheric refraction, radio waves can be propagated a little further, even in normal temperature conditions. Due to this phenomenon, the "radioelectric Earth's radius" can be estimated as 8,500 km. Thus, the effective direct range of a 300 meter tower is about 70 km, instead of 60 km.

We should notice that radio waves comply with the physical laws of any wave such as reflection, diffraction and refraction: to be propagated in "free space conditions", they need to be free of any obstacle not only on the straight path between the transmitter and the receiver but in a global volume, called the Fresnel ellipsoid. At the middle of a path whose length is L, a wave whose wavelength is λ needs to be free of any obstacle (such as the ground), in a volume whose diameter is 2 hf, with

$$2 \text{ hf} = (\lambda \cdot \text{d})^{\frac{1}{2}}$$

As an example, a link using a 100 MHz carrier, on a 30 km range, needs to be set on a line-of-sight path which is 150 m above the ground. For a microwave radio link at a 4 GHz frequency, over a 50 km distance, the free space is only reached at 30 m above the ground.

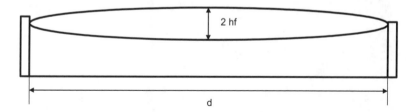

2 hf

d

Figure 3.3. *Fresnel ellipsoid. Note: heights are greatly enlarged. For near free space propagation conditions, not only the line of sight but a full volume, called the Fresnel ellipsoid, must be free of any obstacle. This implies that the line of sight path between the transmitting and receiving antennae be at a minimal height "hf" above the ground*

It is not mandatory that the Fresnel ellipsoid be fully free of any obstacle for efficient communication. However, in such conditions, parasitic phenomena may happen, such as reflections on the ground, which can disturb the propagation and lower the transmission range. It is not necessary to get deeper into such situations but only to stress the importance of choosing appropriate sites to install radio equipment and to comply with some magnitude orders for the geometric parameters of transmission paths:

– for long distance terrestrial links using waves in the metric or decimetric bands, it is often necessary to choose natural summits or man-made high towers to achieve free-space conditions. This is typical of TV broadcasting;

– for centimetric point to point links, operators commonly use towers whose height is about 100 meters, placed at a distance of 50 km one from another;

– for millimetric links whose range is naturally limited by atmospheric losses under a few kilometers, most installations can be made nearly anywhere, on very small masts if necessary, and the line-of-sight clearance is sufficient for good propagation conditions.

These considerations explain some features of the "radioelectrical landscape". The valuable high sites are a scarce resource and it is not possible to build high masts or towers anywhere, especially in towns, for environmental and urbanism reasons. This is why radio systems must regroup at particular places. Local regulation may be necessary to organize and manage such radio areas where very different equipment needs to be installed. State-of-the-art engineering is necessary to avoid any interference problems between them (see Chapter 2, section 2.5).

Figure 3.4. *A common radio site. The radio stations group themselves on interesting or remarkable high sites which are well adapted to wave propagation. Local regulations may be needed to facilitate and organize such common places which are shared by a number of systems and operators often using the same mast infrastructures. Such places can be usefully managed by a site operator which is chosen to enforce and control good radio engineering practices. Periodic inspections may be appropriate*

In the same manner, it should be pointed out that some sites must be protected for good reception of very low radio signals. This is the case for some satellite earth stations. The most accurate case is radio astronomy where extraordinarily weak signals are searched for from space. Referring to Chapter 2, the signal quality is determined by the C/N ratio between the useful signal power C and the noise power N which enters the demodulator. If the C value is nearly independent of the chosen terrestrial site when the radio source is in space, the noise power N can be very dependent on the radio environment. Thus, it is a crucial concern, when the link is so sensitive, to find a protected receiving area where the radio noise, which can be considered as a true pollution by such services, is at a minimum. It can be some natural depression or valley, in remote countryside, where surrounding heights provide a screen for the radio signals coming from elsewhere, notably from towns. Such sites are becoming more and more scarce too. They can be found in mountains or inside old volcano craters, but with more difficulty in ordinary regions. It may be suggested that special regulations be locally enforced to protect also such places which are absolutely necessary for some radio services (see also section 3.7). Along the same lines, someone once suggested that the hidden face of the moon would be an ideal place for radio astronomy since the Earth satellite would protect the receivers from any man-made signal.

3.3. Propagation on the Earth

The free space conditions are partly theoretical and the equations, as given above, can only give a rough idea of the reality. For space services such as satellite communications, these mathematical models fit pretty well with the observed facts and are then acceptable with minor corrections. However, for a number of terrestrial services, the free space conditions cannot be obtained. Some phenomena such as ionosphere reflection have already been mentioned for decametric waves but many other factors can disturb the wave propagation along the ground surface, through the atmosphere. Let us mention mechanisms such as reflection, diffraction, refraction, scattering or attenuation by hydrometeors.

Some mechanisms increase the propagation range. This is the case of ionosphere reflections or scattering. The same is true for refraction in the lower atmospheric layers (see section 3.2). In some atmospheric conditions, when a thermic configuration is appropriate, especially over the sea, remarkable propagation ranges can be observed due to wave ducts which drive the waves for hundreds of kilometers. This phenomenon can be compared to optical mirages. Diffraction may also be common and can sometimes be intentionally used. As an example, it may be observed on a mountain crest or along a roof ridge where the waves can be strengthened in a particular direction, making possible a connection with out of sight stations.

Other mechanisms weaken the waves. In most cases a network engineer tries to keep clear of evident direct obstacles along the propagation straight line. It may be more difficult for him to take care of multiple paths. If a wave follows two different paths, the combined radio signal shows interferences, according to the same principle as optical interferences, with interference fringes spread over the field. At some places both waves are phased in and the signal is there at a maximum, but at some other places they are phased out and nearly no signal exists. The distance between such extrema is half a wavelength. This interference situation is common and can be met in very different contexts where reflections take place. All frequency bands are concerned.

It is also necessary to cope with atmospheric absorption and rain falls. Here, only the waves at a high frequency, typically microwaves, are concerned. Depending on the wave frequency and the rain density, losses are more or less significant but may sometimes be so high that the link is interrupted. However, such disturbances are temporary. On the contrary, the Earth's atmosphere is permanently opaque to particular radio waves, such as those in the 60 GHz band, oxygen absorption band. Waves in this band can however be used for very short distance links with the advantage of an excellent protection between systems using the same frequencies since the atmosphere is an almost perfect screen.

It should be stressed that, in many circumstances, due to propagation conditions, the received signal power fluctuates, sometimes in huge proportions, along time scales which may also be very different. This fact explains why the propagation attenuation Ap (see Chapter 1) may be difficult to calculate. It is very often a statistically defined parameter. As an example, at some place, the radioelectric field intensity value may be guaranteed for 99.9% of time. Its value for 99.99% of time will be lower. Depending on phenomena, losses may be long lasting or very short term thus the quality objectives should be set along different time scales: a year, an hour, a second. Technical strategies to cope with such statistical losses are of course very different: from site diversity to digital protocols and error correcting codes.

Some simple technical devices such as space diversity can partly cure problems such as multiple path interferences with different antennae placed at a distance from one another and used simultaneously. Frequency diversity can be preferred or added when the losses are frequency dependent, using two or more different carriers for a single message (this is common for HF ionosphere communications). However, this technique is spectrum consuming and should be reserved for security communications. Nowadays, digital processing provides new powerful tools for reliable reception of fluctuating signals. Some of them are on-line procedures, such as frame transport protocols and error correcting codes. Some, such as MIMO (multiple input-multiple output), are sophisticated modulation schemes where different radio carriers are combined which are generated by different message synchronized sources.

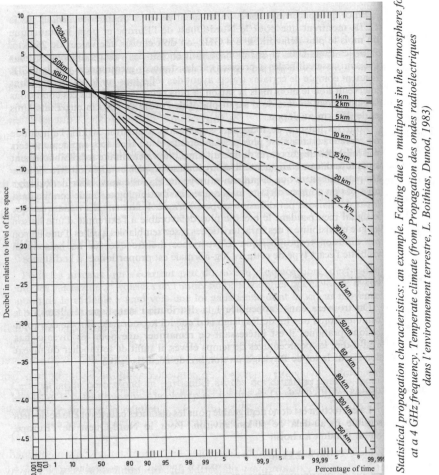

Figure 3.5. *Statistical propagation characteristics: an example. Fading due to multipaths in the atmosphere for waves at a 4 GHz frequency. Temperate climate (from Propagation des ondes radioélectriques dans l'environnement terrestre, L. Boithias, Dunod, 1983)*

Mobile phone services have encountered particular propagation conditions in urban areas where free space is nearly an exception, due to the numerous buildings and reflections of all kinds. In towns, the field statistical distribution is very irregular and the radio field intensity for mobile services varies rapidly and deeply when a mobile is moving. This situation, called Rayleigh fading, is characterized by a rapidly fluctuating signal at the receiver input. The radio network engineering and the digital modulation and protocols used must adapt to such a context.

A number of mathematical models have been developed to describe the electromagnetic field statistics in various environments.

A radio network specification should take into account the required service performance but also the propagation statistics and more generally all the features describing the geographical context where the network will be implemented, in order to optimize its engineering, also keeping in mind spectrum efficiency. Fortunately, digital technologies which offer capacities for signal processing and communication protocols are able to deal with some propagation constraints which would be unusable using analog modulations. In any case it is necessary, to achieve a good technical design, to get context models for our disposal which describe the propagation environment as well as possible. A close adjustment of networks to complex radioelectrical and geographical conditions is a mark of modern radio engineering.

3.4. Space, orbits, satellite systems

Space radiocommunications were born in 1957 with the "Sputnik" satellite and the famous "beep…beep…" from its radio beacon which was broadcast all over the world. The only means to communicate with space and a satellite is a radio wave. Thus, a new era began for radio. Every satellite is provided with different telecommunications devices in order to keep connected with the Earth: uplinks from Earth to satellite, downlinks from satellite to Earth, ancillary links for telemetry or technical control. Today, intersatellite links are also possible. All of them use different frequency bands which have been allocated for these different cases.

Among their various applications, satellites or "space stations" as mentioned in the Regulations, can be extraordinary platforms for communication between any two points on the Earth, the satellite being a relay for the radio waves transmitted and received by "earth stations" which are directly in view of this platform. This was clearly demonstrated by Telstar (1962), Early Bird (1965) and others, such as Molnya (1965). Since then, satellites have been a major technology for long distance communication, especially internationally and intercontinentally, and have completely replaced HF links, at least for commercial purposes. For some time, satellites also had a monopoly on long distance TV transmissions. Now they play a major role for direct TV broadcasting to customers. In some domains such as radio navigation, environment studies, Earth observation, they are essential. Space communication management will be later discussed. Now, only some considerations are introduced about the geographical constraints of satellite communication.

Satellites move on orbits around the Earth. Many different kinds of orbits are possible, however they all follow Kepler's laws. They are ellipses, one focus being the center of the Earth, and are run according to the area law which states that, considering a radius from the satellite to the center of the Earth, it sweeps equal areas in an equal time. As an example, if a satellite runs on a very elliptical orbit, it stays much longer around the apogee than the perigee. A satellite is "seen" from any point on the Earth as describing a trajectory in the sky, depending on its orbit. Its position can be calculated at any time with great precision and is described by its azimuth and elevation.

From the spectrum management point of view, two main categories of satellites should be distinguished: geostationary (geo, as an abbreviation) and non-geostationary (non-geo) satellites.

Geostationary satellites move on a circular orbit in the equatorial plane, at an altitude of about 36,000 km. On such an orbit, they rotate around the Earth with a 24 hour period, exactly synchronous with the Earth's own rotation. Thus, they are seen as fixed in the sky. They can be used as a permanent radio relay, as if they were on the top of a huge tower. However, such orbits cannot be perfect and geostationary satellites may slightly move around their nominal position. This movement should be taken into account when very directive antennae are used to communicate. However, it may be ignored by antennae with a low or medium directivity such as direct TV antennae which see the satellite in a reasonably fixed position.

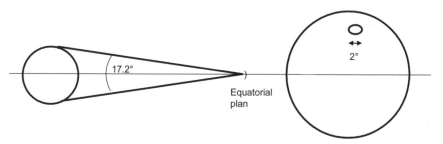

Figure 3.6. *The Earth, as seen from the geostationary satellite orbit*

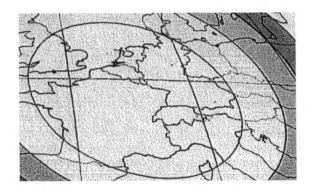

Figure 3.7. *Coverage of a 2° spot antenna over Western Europe*

From a geostationary satellite, at an altitude of 36,000 km, the Earth is seen at a 17° angle. "Global beam" antennae which are designed to cover the greatest possible terrestrial area should provide a radiation pattern which fits with this aperture angle, a little narrower, in fact, not to illuminate the Earth with low elevation waves which could impair terrestrial radio links in the same frequency band.

To cover a particular country, or region, the satellite is equipped with "spot beam" antennae. Figure 3.7 shows the coverage obtained from a 2° spot beam antenna over Western Europe. The spot beam antennae have a greater diameter than the global ones: the more accurate they are, the larger the satellite should be.

The geostationary orbit (the right regulatory wording is "geostationary satellite orbit") can be shared by a great number of satellites. Of course, if they use different frequency bands, they can be placed side by side without any interference problem. In this way, most satellites carry several payloads, with equipment using different frequency bands. However, if they use the same bands, two satellites should be placed on such nominal positions that they do not interfere with each other for any terrestrial station. It is also necessary to use antennae with an appropriate directivity, in order not to mix signals.

From a station on the Earth, the geostationary orbit is seen as a bow in the sky, stretched from one point of the horizon to another, which culminates at an elevation depending on the latitude. From Paris (latitude 48°N), a geostationary satellite positioned about the same longitude is seen at an elevation of 35°. At an equatorial place, the geostationary orbit crosses its zenith. The geo orbit cannot be seen at latitudes higher than 81.5°.

Let us imagine different satellites placed along this bow with an angular distance of 3° and a position tolerance greater than 1°. A terrestrial station with an antenna whose directivity is greater than 1° will be able to communicate with one of these satellites without interfering with others. Thus it is possible to plan the geostationary orbit with longitudinal positions as a frequency band is planned with channels. This combined planning of orbital positions and spectrum bands is a major achievement for ITU, since this couple of resources is strategic for satellite communications.

From a geostationary satellite, only part of the Earth can be seen which is a spherical portion centered on the equator with a longitudinal extension of about 160° and a latitudinal extension of 80°N-80°S. Polar regions cannot be covered by geostationary satellites. The useful terrestrial area is in fact a little smaller because, as mentioned above, earth stations should work at a minimum elevation (typically above 5°) and the satellite should not illuminate the Earth with almost horizontal waves. Under such conditions, three satellites are necessary to cover the 360° of equatorial longitude and the main part of the continents. It is a major advantage of space communications that, within the areas where a satellite can be seen at a minimum elevation, it can be accessed from any point on the ground. Thus, as soon as such a satellite is placed in orbit, a universal service is available.

A number of satellites are not concerned with worldwide coverage but should only illuminate some specific service areas. Thus the same frequencies can be reused on different spot beams. According to their antenna radiation patterns, satellites may cover some precise regions on Earth. Positioning exactly and maintaining the spacecraft attitude are of major importance to keep their service area pattern on the ground stable. Thus spectrum efficiency, in space, requires that greater and greater satellites be launched, with directive antennae directed with an ever growing accuracy. A local coverage adapted to a middle sized country or a region is common.

Non-geostationary satellites can be placed on various orbits, depending on their service objectives. These orbits are generally at a lower altitude than geostationary satellites, the satellite period being thus shorter than a day. However, some systems use elliptical orbits with an apogee very far from Earth. The common feature of non-geostationary satellites is that they appear to move in the sky, staying in view of any station only for some time. Thus, to insure a permanent service, a cluster of satellites must be used with such orbits so that, at any time, at least one satellite is seen from any customer and several for some services such as radiolocation. Such configurations may imply that communications are switched from one satellite to another and make network management more sophisticated. Earth stations, if they use directive antennae, must also track the moving satellite.

Non-geostationary satellites offer some advantages:

– they can cover any place on Earth;

– if they are placed on an appropriate elliptical orbit, they can stay in a nearly fixed position for a long time by the zenith of a service area, which is an interesting situation for broadcasting, as an example;

– on the other hand, if they are running on a lower orbit than geostationary satellites, the path attenuation is less and smaller terminals with a low directivity can be used, such as mobile telephones or beacons;

– the transit time for communication through a low orbit satellite is also less since the transmission path is shorter. This transit time is about 250 ms for communications through a geostationary satellite (about 75,000 km to be traveled by a 300,000 km/s wave);

– when the service concerns terrestrial survey, low orbit satellites can perform more accurately for observing phenomena.

Managing the spectrum-orbit resource for non-geostationary satellites is more complex than for geo. The orbits should be interlaced when different systems share a common frequency band and the satellite activity periods should be strictly controlled to avoid interferences. Several abbreviations are used to designate different satellite systems according to their orbit type, such as:

– GEO, geostationary satellite;

– LEO, low orbit satellites;

– MEO, middle orbit satellites (intermediate between GEO and LEO).

Let us point out to end this section the extraordinary precision of calculations, based on space mechanics, to determine a satellite position at any moment. The decimeter order of magnitude is achieved when necessary, for precise radiolocation as an example, or to make interference measurements in space.

3.5. Terrestrial network coverage

Satellites, as mentioned above, instantly provide a wide service area. The same cannot be said for ground-based networks where any station can only provide a limited range of coverage, due to propagation and geographical considerations. Thus it is necessary to deal with the case where a broad network, as large as necessary, has to be built from individual ground stations, with a limited spectrum. This objective can be achieved by an appropriate engineering of stations, reusing frequencies in the most efficient way. Planning the spectrum with standardized channels is a powerful tool in this context.

The problem to consider is interference between radio links. If a transmitting station, using a particular frequency, offers a limited range of coverage, it can be imagined that, in a far enough area, the same frequency is reused by a different station. This frequency reuse does not create any problems if an appropriate propagation protection exists between the two areas. It means that, at any point, the signal amplitude received from the nominal transmitting station is much higher than the signal received from another station using the same frequency. On the other hand, if waves are received somewhere from different stations at the same frequency and with nearly equal intensity, it may be impossible to make a distinction between them: an interference occurs.

Thus planning strategies have to be imagined to build large networks from a limited spectrum, which make it necessary to reuse the frequencies as much as possible without interferences. As mentioned above, channelization is the most common practice and typical channel reuse schemes have been designed for different situations, such as the following:

– long distance transmission. Let us suppose that a message has to be carried a very long distance by radio links installed on towers built on sites regularly distant from one another. There, the message is relayed by radio stations equipped with very directive antennae. In common circumstances such links are bilateral and symmetric. As for any duplex system (see Chapter 2, section 2.4), two frequency bands are generally used: a higher frequency band and a lower frequency band. They are planned in parallel and divided into standard channels. A long distance link can be realized with a single channel couple, one channel from the higher band and the paired channel of the lower band. Each channel of this couple is alternatively transmitted in each direction. Cross polarization of waves (see Chapter 1, section 1.1) is generally used to improve the mutual protection of sites;

– broadcasting. Here the objective is to provide a unique message uniformly, such as a TV or radio program, to the greatest number of receivers spread over a large area. The link is unilateral. Generally a main transmitter, installed on a high site with a large aperture antenna, illuminates the widest surface, using a particular channel of the frequency plan. To enlarge the service zone on remote locations, secondary transmitters can be installed beyond the main broadcasting area. However, they frequently use other channels from the spectrum plan not to interfere with the main transmitter. According to the coverage extension and density, with analog modulations, up to six channels may be necessary to broadcast a unique program. Synchronous relays do also exist and can be used for frequency modulated or digitally modulated waves, using the same channel, but should be strictly synchronized;

– cellular networks. Mobile phones have popularized cellular radio networks. They are telecommunications networks, spread over very large areas such as a whole country thanks to "radio cells" which are placed side by side in order to regularly "tile" the country. The cells are radio independent networks using different channels from a spectrum plan. A frequency reuse scheme is chosen which follows a systematic pattern, such as two contiguous cells never use the same channel and thus cannot interfere. This configuration gives the best opportunity for a denser frequency reuse, when necessary. The network is made denser by making the cells smaller, which means lowering the transmitter powers and relay station heights and increasing the cell number. Such network strategies have proved to be very effective for public mobile telephony, in GSM networks, for example, affording remarkable spectrum efficiency.

Two consequences follow. At first, an increase of the fixed transmitter number obliges the operators to look for more and more transmission stations and to multiply the radio high sites, especially in towns where it becomes difficult to find appropriate places. Secondly, an increase of the "hand over" procedure happens. This procedure combines techniques from radio and computer controlled switches where messages are instantly redirected from one station to another, from one cell to another, when the customer is moving, even on a short distance.

It can be stressed that micro-cellular engineering, with cells covering a distance of 100 meters, or even less, makes it necessary to "lower the high sites" in order to avoid transmit sites which would provide propagation ranges which are too large, thus being an obstacle to an intensive spectrum reuse.

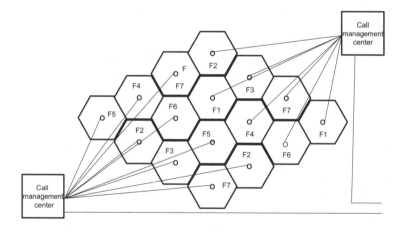

Figure 3.8. *A cellular network. A cellular mobile telephone network is made of radio cells placed side by side, according to a regular frequency scheme. Thus, channels can be reused as much as necessary without any interference between two adjacent cells (here a seven frequency based scheme is shown). When staying in this area, a mobile is located and connected at any time to the nearest radio station. This station has to hand over to another station (hand-over procedure) when the mobile moves to another cell. Switching messages in real time is carried out by remote computer controlled switching centers*

– local networks. A local network, as its name suggests, is a network with limited coverage, typically the coverage of a unique master station: home, company premises, town district, etc. From a regulatory point of view, such networks are often license free and without any particular frequency allocation. Thus it is necessary to anticipate a situation where, by chance, two local networks would be placed side by side and could interfere. To avoid such problems, equipment is designed with some self-adaptive strategies derived from the different multiple access techniques (see Chapter 2, section 2.4), such as an automatic search for a free channel among a set of authorized ones, or a free CDMA code among the orthogonal family. Of course this self-adaptive capability of local networks, if it exists, can be voluntarily used to create a wide common service area from different independent equipment.

3.6. Coverage strategies

If a satellite system can cover a whole country or a continent at once, terrestrial networks for a public service such as mobile phones or TV broadcasting spread themselves in a progressive manner and cannot afford a global service for some time. Investment must follow the market prospects. At first they are devoted to business centers, communication lines and main cities before moving on to less dynamic zones.

However, waves, as public goods, should not be monopolized by the richest parts of a country, even for the most profitable services. Regulators and administrations may concern themselves with a "universal service" with global coverage. On the other hand, operators may be more interested by a "skimming" strategy where investments for radio stations are concentrated or limited to the most active areas. A compromise has to be found which may be included in the operator's license or frequency administrative assignment. As an example it may enforce some coverage objectives expressed as a population percentage to be served year after year.

Often a new terrestrial network spreads from spots centered on the main cities, which grow progressively and later provide continuous coverage of large areas of the countryside. This economical strategy can be backed by a frequency strategy adapted to service zones which differ notably by their traffic density. Many examples can be given of systems concurrently using two or more frequency bands. A higher frequency band can be better adapted to urban centers with short distance links and a high traffic density, whereas a lower frequency band is adapted to countryside or less dense areas. This scheme can still be strengthened by a specific coverage of very high density places ("hot spots") such as railway stations, airports, congress halls, etc. where the traffic reaches summits and where dedicated technologies may be appropriate. Such frequency strategies can make use of "multiband" terminals which combine the different network technologies for a unique service. A good example is the GSM mobile phone which is adapted to both 900 and 1,800 MHz.

3.7. Radio site protection

If a radio station or link is built and works in full compliance with regulations, according to state-of-the-art principles, it may be useful to protect it firmly against any disturbance, mainly but not only interferences, typically when public security considerations are concerned. The Radio Regulations give rights and protection on the basic principle "first came, first served" which may not be sufficient for some cases. The international protection applies to radio stations for preventing mutual interferences, it may not be appropriate to other disturbances and contexts.

Two typical situations are encountered. Radio interference can be created by industrial equipment which is not fully covered by the Radio Regulations, such as electrical stations, and which may generate spurious signals. Propagation disturbances can follow the erection of a physical obstacle along the propagation path.

Due to these possible disturbances, some national laws provide a particular radio site protection which applies mainly to important sites where security reasons or particular public concerns are relevant. They may limit the possibility of installing some electric or electronic equipment unless at a minimum distance from the radio station receivers. They may maintain the past clearance in the Fresnel zone (see section 3.2). Of course such protection must be restricted and cannot be requested for any radio system.

As a conclusion, a radioelectric space does exist, however invisible, which is organized as a cadastre register, following many geographical constraints. The use of frequencies and sites is limited by rules which found the rights and duties of the different operators and stations. However, this organization changes rapidly: rights should be revocable in time and nobody may "own" the spectrum. This is a major difference with land property which may be indefinite. Spectrum management is simultaneously a registration of rights to use frequencies with protection of this usage, and a permanent work to evaluate the existing rights, to limit or dismiss them in order to improve the global spectrum efficiency.

Chapter 4

Spectrum Sharing, Bases and Actors

Managing the radio spectrum means getting the best efficiency from this natural resource which is available everywhere, at any time, for everybody. The game is to temporarily provide every radio service operator with the spectrum resource (generally a frequency band) which is necessary for his service, over the area where he wants to provide it. The service quality to be achieved must be specified to determine as exactly as possible the assigned spectrum share. This spectrum procurement is associated with rules and dispositions which determine the operator's rights and duties.

The radio spectrum offers an indefinite capacity which can always be increased. It is never worn out and is always available. However, a wave can only be used if it is not jammed by others at the receiving end. It should not be transmitted if it may interfere with existing ones. From the beginning of wireless services, all the efforts of spectrum managers have been dedicated to defining with an ever growing accuracy, the conditions of using waves. This effort is necessary to increase as much as possible the number, variety and density of radio systems and to improve the efficiency of the spectrum as a shared universal resource to carry information.

4.1. Radio frequencies: common goods

Who owns the spectrum? This question has no answer and almost no sense. There are so many users and uses in different economical and political contexts that a unique answer is wrong. Evidently, the spectrum is an abstract and mathematical entity, immaterial. This ever renewed and growing resource results from the regulator and engineer efforts to plan and organize electromagnetic fields. Their combined activity cannot be anyone's property. Without their efforts, radio would not be as efficient as it is and would not really be usable.

It can be considered that radio waves, natural or man-made, are a universal phenomenon which experts should control at any place where they want to develop communication services. Thus, the right question is: who controls the radio waves? Who manages the spectrum, here and now?

Practically, on the basis of experience, the radio spectrum is managed by the state governments, according to the Radio Regulations (RR), an international treaty which they formally approved. The International Telecommunication Union (ITU) is the international forum which has been created and chosen to elaborate these regulations and whose Members are the state administrations.

The ITU constitution preamble puts things clearly:

> While fully recognizing the sovereign right of each state to regulate its telecommunications and having regard to the growing importance of telecommunications... the States party to this constitution... have agreed as follows...

Thus, it means that ITU activities are a collaboration of states for the benefit of all people in the world. Spectrum management, as decided in the Regulations, is a fruit of this international cooperation.

Such an agreement is guided by evidence: no state can make use of radio on its territory without any rule and control. Waves ignore border lines and the rights of any state are limited by those of others. Moreover, radio is perfectly adapted to promote international services and thus should be managed from an international basis.

Some radio services need international agreements and collaborations, thus they are stated as such in the Regulations. This is typically the case of communications by HF waves which can be propagated throughout the world, or the satellite systems. However, for many other applications, even if it is not absolutely necessary from a technical point of view, states very often look to ITU agreements, or at least regional agreements, concerning system standards, frequency plans, procedures and any other matters which harmonize the technical and operational features of services. In all cases the main partners in such debates are the states: if the ITU or regional expert bodies play a major role to achieve agreements, they act as "catalysts" rather than deciders.

Often too, the spectrum resource (and satellite orbits) must be shared between different countries which compete for them. This sharing is organized by the Regulations or regional agreements. Two common practices are the following:

– either the sharing is decided, from the beginning, by an international "allotment" plan;

– or frequencies and channels are attributed according to the principle "first come, first served" without prejudice of former rights.

Both ways are proposed by the Regulations depending on the frequency bands and services.

In addition to situations where international coordination or spectrum allotments take place from the beginning, states have the full responsibility of managing the whole radio resource on their territory, in compliance with the Regulations. However, they find it an advantage to establish bilateral agreements with their neighboring countries to harmonize their standards and practices for a better use of frequencies in the border zones. Such agreements are called cross-border coordination and are useful to minimize interferences, especially when the frequencies are requested according to the principle "first come, first served".

When a new radio station should be installed, a procedure takes place under the responsibility of the concerned administration to care about interferences. This procedure is fully national if the station is far from the border and has no impact on any foreign service. Or it is ruled by the cross-border coordination agreement if foreign stations are possibly impacted. No implication of the ITU is needed but, at the end, the international organization may be requested to register the station which has been approved, with rights granted by the parties. This way of doing things is very common and makes direct involvement of the ITU exceptional.

To sum up, national procedures, even when harmonized with neighboring countries, are applicable on each state territory. They should conform to the Regulations and have regard to the equality of rights of every state, especially in border zones. ITU can be a referee for any interpretation of the Regulations.

In any case operator rights on the spectrum are given by the administration of the country where the station is installed. These rights may be founded on purely national bases, or may have been negotiated with neighboring countries or within the international community at the ITU. In such cases they have been obtained by the administration through regulatory procedures and they are the administration's property. It is an administration privilege to commit such rights to an operator for its benefit, always under the Regulations' constraints.

This monopoly of states to issue rights for the spectrum usage by private or public operators, even when some frequency bands are said to be "license free", may suggest that states are the spectrum owners on their territory. However, this "ownership" is greatly dependent on international rules and constraints and should be better considered as spectrum "control".

A state government has the responsibility of making the spectrum resource available to users and customers in its territory. It does so according to the needs which are expressed by manufacturers and operators, with its own choices of priorities and procedures, due to a limited resource. It may request fees and taxes for this use of frequencies and impose technical and administrative constraints which are stated in a license or as a general law. For other countries it is responsible to make the different operators, being commercial or administrative as well, comply with the Regulations and any international agreement.

Due to these monopoly, constraints and responsibilities which frame this state "ownership" or "control" of the radio spectrum on their territory, it is obvious that any frequency license, as issued to users, should be time limited, or at least a termination clause should be included. No administration can estimate to manage an indefinitely available part of the spectrum. It cannot guarantee to any user the indefinite benefit of any frequency band. An international negotiation may always modify the Radio Regulations and change any band allocation which is no longer available for some existing services. This precariousness of the spectrum usage, which will increase with the radio service growth, should always be considered when debates take place about spectrum value and its management as market goods.

4.2. Regulatory services for spectrum sharing by the ITU (allocation frequency tables)

The ITU organizes the general sharing of the radio spectrum. The whole spectrum is allocated to services by the Radio Regulations, at least the most useful part. In fact, in the field, it is more or less used, with variable effectiveness and intensity, depending on the countries and geographical areas. However, no spectrum, up to 300 GHz, is left without any, at least theoretical, regulatory allocation. Each band is allocated to one or more "services", either worldwide or depending on Regions. Only these services should be implemented in this band. It does not mean that such services exist or are implemented: the spectrum allocation is made *a priori*, on a technical basis (see also Chapter 7).

The word "service" has been written above with quotation marks to stress the fact that it has here only a regulatory value. The services, as designated by the Regulations, are several broad radio application categories which are described by global purposes or features and major operational constraints. Thus, particular rules, procedures and technical specifications can be issued about each of them to adjust, as exactly as possible, the radio resource to their general needs and characteristics. As examples, some of the most important regulatory services are the following:

– the terrestrial fixed service which concerns permanent links between fixed and determined points on the Earth;

– the terrestrial mobile service for temporary bilateral links with mobiles whose position on the Earth is undetermined, mainly from a fixed station;

– the terrestrial broadcasting service which links a center station to receivers with an undetermined position, for unilateral programs.

Similar services can be offered through satellites. Each corresponding service is no longer "terrestrial" but "satellite":

– fixed satellite service;

– mobile satellite service;

– broadcasting satellite service, etc.

Roughly speaking, between 50 and 100 different services are distinguished. Chapter 5 describes some important features of these main services as understood by the Regulations. It should be pointed out that they are not specifications or technical descriptions of real products for manufacturers and customers. They are theoretical tools oriented for the benefit of spectrum management and categories among which any real radio application or system has to find its place and the associated regulations. A common question when a new radio product is being imagined is: "which ITU service does it correspond to?" The answer helps to find which frequency bands can be used.

Thus, the ITU Regulations do not consider practical applications or actual services as we all know in our modern life. As an example, a public mobile telephone service, such as a GSM network, is a terrestrial mobile service, as is a fire brigade radiotelephone network. Local private company wireless networks, RLANs, are also mobile services. On the other hand, FM radio and terrestrial TV are broadcasting services. Long distance infrastructure radio links are a fixed service.

The services, as defined by the ITU, are used to build rules and procedures which fit as well as possible with the corresponding applications. The objective is to give them the necessary radio resources but with limits in order not to impair rights already existing for other services. In fact, a fair balance has to be found between the different services in order that they can all grow according to the social and economical needs, but with acceptable restrictions and constraints. Techniques and markets change so rapidly and so much that it would be impossible to adapt the regulations for any new application or prospect. Thus, it is necessary for the ITU to consider only a limited number of services, as large families which have to share the resource on the basis of their general features, without considering the family members. The number of combinations therefore remains limited and the spectrum management should not deal with indefinite particular cases.

In every band a distinction is made between primary and secondary services. The authorized primary services have priority over the secondary services which share the same frequency band. This means that a secondary service cannot complain if it is disturbed by a primary service in the same band. In addition, a secondary service should not interfere with a primary service. Of course the choices of primary and secondary services which are authorized in a particular band have to insure the best possible mutual compatibility in order that they can work simultaneously in the most common circumstances. Some priorities may also be decided between primary services.

Having set these basic principles, the spectrum is divided into bands or frequency slices, more or less wide, which are allocated to different services in a frequency table. Some bands in the millimetric wavelength domain are global blocks, several gigahertz wide. On the other hand, in lower frequency domains, the bands may be much narrower and allocations detailed with a precision of a megahertz or even a kilohertz. This difference depends on the needs for some frequency bands, more or less usable, but also on the necessity to implement some worldwide particular services which should be specified with the highest precision, such as rescue or security services and some reference radio beacons.

On the other hand, if the propagation range of services is rather low and if no international common objective is clearly defined, the management of corresponding bands can be left to national administrations. Thus, the ITU keeps the regulation broadly open and does not issue detailed recommendations.

4.3. The role of states in sharing the spectrum

In conformance with the Radio Regulations and taking care of any regional agreements and conventions, being respectful of allotment plans between neighboring countries, if they exist, a state government and its administration are responsible for the spectrum resource in their territory and should make it available to users and operators for practical radio applications.

For this objective, they have, at first and according to their responsibilities and priorities, to define more exactly the spectrum organization and allocation tables.

They also have to issue more detailed administrative and technical rules than those stated in the Regulations and to make them compulsory. They have to decide on any provision which helps to improve the spectrum use for a better efficiency and effectiveness in their country.

Let us consider, as an example, a band which is allocated by the ITU to the terrestrial fixed service. It is the privilege of a national government to share this band between its Defense ministry for military needs and a civil independent authority for commercial operators. Thus, these bodies can decide more specific uses under their own responsibility such as long distance infrastructure links, fixed customer termination links, etc. They may also specify a detailed frequency plan with precise channels and choose standards for the equipment which will induce a sub-channel organization, etc. A part of the spectrum may be managed on a nationwide basis, another left to local and regional authorities.

All decisions, as illustrated above, are relevant to the state administration and any spectrum problem which would happen from the radio usage, in this country, has to be dealt with by the administration, particularly in relation to foreign countries, in view of the Regulations provisions.

As a common practice, the national rules and spectrum allocation tables, at least the main ones, are published as official administrative documents. Most countries issue a national frequency allocation (or utilization) table which is publicly available. Besides this general basic document which can be easily obtained, more detailed channel or frequency tables can be found, directed towards particular applications. Other professional tools may also be proposed to operators which are useful for satisfying spectrum management procedures and controlling the frequency resource.

The most evident and simple administrative practice to conserve the spectrum is to limit the bands allocated to every service and the channels assigned to every operator strictly. Such a policy leads to an ever growing planning and division of the spectrum since the number of services and operators is ever increasing. The radio spectrum becomes divided into bands, sub-bands, channels and individual carriers. This general practice of "slicing up the spectrum" is comforted by the current international standardization policy where the chosen radio modulations are frequently adapted to such a frequency detailed planning. However, this prevailing policy has been recently challenged to keep bands open to wideband or "spread spectrum" applications (see Chapter 10).

A second way to conserve the resource is geographical containment. The channels are assigned on a limited area: the nation or a region, town, single transmission site, with some technical features in order that waves remain enclosed in a limited area or, when propagated away notwithstanding administrative and border lines, do not interfere elsewhere.

Such technical specifications, in line with international or purely national standards, mainly concern the transmitting stations, to limit possible interference with other users:

– maximum power or EIRP;

– restricted radiation antenna pattern;

– restricted spurious and out-of-band signals;

– geographic site exclusion;

– time cycle use, etc.

Sometimes, receiving station standards may also be requested.

All these administrative and technical tools are parts of national regulations when they are not included in the Radio Regulations. They are enforced as public regulations or specified as particular provisions in the frequency utilization license of operators. However, such national or individual specifications should not be an obstacle to the concurrence equity and the harmonization objectives. International bodies, such as the European Union Commission, are concerned with them to keep the market open and to maintain fair competition.

Frequency band	Power	Channel spacing	Duty cycle (%)
433.050-434.790 MHz	10 mW e.r.p.	No channel spacing	Below 10%
433.050-434.790 MHz	10 mW e.r.p. –13 dBm/10 kHz for wideband channels	No channel spacing	No duty cycle restriction
434.040-434.790 MHz	10 mW e.r.p.	Up to 25 kHz	No duty cycle restriction

Figure 4.1. *Technical limits chosen in Europe for low power devices (LPD) around 434 MHz (from the CEPT decision ECC/DEC/ (04) 02). The radio systems may be subject to particular technical specifications in order to limit the possible interferences with other systems which share the same frequency bands. Such limits concern mainly the transmitted power or antenna radiation pattern. Here is an example from the CEPT decisions concerning low power equipment such as telemetry, remote controls, alarms, etc. in a dedicated small band around 434 MHz. It can be noted that the decision also gives indications on recommended receiver specifications*

4.4. How to plan new applications and compatible services

When a new radio application is imagined, a question follows: which frequency band can be used?

– Is an empty band available?

– Is it possible to withdraw a former application from a band and reuse it for the new one?

– Is it possible to add a new application in a band where another one is at work, both of them being compatible?

Of course the spectrum planners generally encounter the third situation since the spectrum is crowded. However, the first two may happen.

Today, free bands are only available in the millimeter and sub-millimeter wavelength domain, where waves can be used, as an example, for short distance applications or, on the contrary, for space communications.

Some opportunities may be found to reuse a band when a recent application had no success and can be given up. There is also the case when an application, even though still at work, is thought to be obsolete and can be abandoned for a new one with a higher priority or benefit. This has typically been the case for recent years, when some long distance infrastructure radio links, replaced by optic fiber cables, have been dismantled to find frequencies for mobile services. However, in such a situation, it may be a hard job to remove the old equipment in order to free the band. The operation should be thoroughly planned and financial incentives may be appropriate to hasten the movement. Administrative and financial tools have been developed and used in some countries, such as France, to make it easier.

However, as mentioned above, for a number of situations, new systems must be introduced in bands which are already used, after the possible interferences have been evaluated between the new and existing equipment. This evaluation is based on their reciprocal electromagnetic compatibility. The new system should work properly among the radio fields, as they are. In the other direction, the new system should not harm any existing application. Thus, an efficient spectrum organization needs to be based on realistic models of networks and services, existing or in prospect.

Such models concern the technical radio features of systems but also the station geographical situations. The key factor for the compatibility assessment is the ratio C/N or C/I (as seen in Chapter 2, section 2.3) which should remain higher than a threshold value for any link, to insure that it works properly. This ratio depends of the received power, C, and the noise, N, or the interfering power, I. These values can only be estimated from the distance and reciprocal orientations of the radio stations. The calculations may be exact or probabilistic, depending on services and the available information on every station geographical position.

Thus, different strategies can be planned, depending on whether the relevant service is allowed or not in a candidate band, according to the ITU Regulations: is this band allocated to such a service?

Should the answer be negative, a change has to be introduced in the Regulations. This can only be done by the World Radiocommunication Conference (WRC) which convenes about every three years. The process is long, awkward and politically influenced. In any case it should be backed by very strong technical arguments showing that security margins are provided not to harm existing applications. Since many services and applications may be concerned, with different interests, it is necessary to demonstrate that all of them are only marginally impacted. If any risk is detected, a decision will probably be postponed since the assembly remains conservative and prudent. A new service allocation may need to specify some technical restrictions to guarantee compatibility: this has been the case, in recent years, in order to make the geo and non-geosatellite systems compatible with each other and with the terrestrial fixed services in the same frequency bands.

If such an international decision is not achieved or appears too slow or heavy to obtain, a country or a group of countries may prefer to depart from the regulation when they are convinced that the new service does not impair the rights of any other country. However, this derogation remains legally weak and may not be maintained if, any day, an application in conformance with the Regulations is impacted by such a derogatory service.

When the Regulations authorize the service, things are easier and depend only on national decisions to implement any corresponding application, provided that the necessary international coordination procedures are satisfied. In that context, the concerned national administrations have to decide on the technical conditions and procedures which will guarantee the electromagnetic compatibility of the actual systems in their territory and across borders. The same evaluations as discussed above (C/I threshold criteria) are the basis of such coordination between neighboring countries.

4.5. Regulation, harmonization, planning

Up to now, a stress was put on international regulations, notably on spectrum allocation to regulatory services. However, it has been explained that such general spectrum planning is only a framework, a skeleton, which shapes the spectrum to the benefit of large application families. As also mentioned, such ITU regulatory services are theoretical categories. Similarly, the Regions, as described by the Radio Regulations, are global geographical entities without any strong unity, where many countries are gathered with very different economical or political profiles.

Thus, it was stressed that the state administrations should be more directive than the ITU in organizing the spectrum in their territory for actual and practical applications to the benefit of operators and users. However, this national spectrum regulation (also the most important part of the ITU regulations) generally proceeds from common goals and prospects which exist between different countries willing to harmonize their radio applications. Time is gone, if ever, when any country could plan the spectrum and invent new radio applications without reference to others. For many reasons, the main choices are joined between partners. Many working parties, forums or formal structures were born to discuss needs, standards, constraints and rules and to set up strategic orientations for different purposes. Some bodies work to harmonize a particular geopolitical area such as Europe, Americas, Arab countries, etc. Some are devoted to a theme: civil aviation (ICAO), maritime communications (IMO). Some combine geopolitical and thematic concerns (NATO). In addition to some purely national initiatives which can only be launched by major countries, most projects are studied by such intermediate parties which issue common proposals, backed by a number of associated experts representing the administrations but also industry, consumers and so on. Thanks to so many participants, the new applications are found credible and attractive by the market and are more easily accepted by regulatory institutions. If any change is required in the Radio Regulations, it can be introduced and adopted in the ITU with the support of a great number of Members which have been interested and associated from the beginning, in view of practical objectives.

This voluntary harmonization is the most appropriate way to give life to the international regulations and to make them effective. Thus, the actual rules in a particular country, for radio applications and spectrum management, result from different but convergent approaches: the Radio Regulations, the harmonization directions, the national arrangements. The three dimensions are important and should be taken into account by operators and manufacturers.

As an example, let us consider the GSM, the most popular mobile telephone system in Europe and in the world during the time around the year 2000, working in frequency bands around 900 MHz:

– the Radio Regulations authorize the terrestrial mobile service between 862 MHz and 960 MHz in Region 1;

– the CEPT (European Conference for Post and Telecommunications), a conference to harmonize radiocommunications in Europe, decides that bands 890-915 and 935-960 MHz should be used for a pan-European cellular digital public mobile telephone network (GSM standard). This decision is approved and reinforced by a directive of the European Union, becoming law for the union members. Bands 880-890 and 925-935 MHz are reserved for extension bands;

– the national administrations divide the band into sub-bands to license different operators as competitors on their market. Each chosen operator obtains frequency blocks covering the whole territory or part of it.

The case clearly shows how the three dimensions of regulation combine for a global spectrum policy.

4.6. Is the spectrum resource scarce?

All efforts, as described above, may lead people to believe that the radio spectrum is a scarce resource and that the main objective is to manage its scarcity. The diversity of working parties and regulators, the long time needed to obtain decisions and the ever growing number of their participants seem to foster the idea that a finite and unchanging resource has to be shared between more and more services and users. However, simultaneously, the users must acknowledge an extraordinary growth of radio systems, being more and more various and powerful, for the benefit of everybody. Who could imagine, 50 years ago, the "mobile phone revolution", the worldwide satellite services and the broadcasting of thousands of TV programs, everywhere? At that time, already, the spectrum was thought to be crowded! Thus, is the radio spectrum scarce or abundant? What is the truth?

In fact, the continuous activity of telecommunications experts to evaluate and model radio services, to assess the spectrum load and usage, when combined with the permanent progress of technology, improves the resource efficiency, year after year. It can be thought that there is no limit to radio services, neither in quantity nor quality. But this improvement is only possible through tighter optimizations, based on a precise knowledge of services and stations, considered as a "radio landscape" which should be mapped exactly. This effort explains why spectrum managers get a growing influence. Their job is to anticipate and evaluate needs, to plan the frequency bands, to set up regulations and make them effective through procedures, to monitor the usages.

Many technologies which frame the modern "information society" are used to improve spectrum efficiency. All of them cannot be listed:

– radio technology, with integrated components which generate and modulate waves at a still higher frequency;

– signal processors or microcomputers which help to operate sophisticated telecommunications protocols, to code information and messages more efficiently, to detect signals in worse conditions;

– big software programs which make networks more flexible and intelligent to switch the communications instantly according to the service needs and decisions.

We could also mention space technology and all tools available in the background: digital geographic maps, propagation models, real-time computer files, high stability clocks, measuring equipment, etc.

The result is to make the allocated resource more exactly adapted to the needs. Thus, the engineer can design and use electromagnetic fields which are strictly suited to the message to be carried and the links to be established. This art could be designated as wave "confinement" or "containment".

The most realistic description of the spectrum status could be a precarious equilibrium between needs and resources. A permanent scarcity combines with a steady progress which makes radio available for more uses, but in controlled conditions. The permanent scarcity is the goad to research new tools and practices for a greater efficiency. At any time the resource is limited but sufficient to reach a balance on the spectrum "market".

4.7. Spectrum sharing: a summary

Having described above in some detail the general principles of spectrum sharing and the main actors in charge of this resource management, it can be useful to sum them up with particular attention paid to the vocabulary which describes the different steps of this sharing: this is part of the Regulations.

The radio spectrum, in each *Region*, is divided into frequency bands with homogenous characteristics which are *allocated* to different *services*. These services are large technical categories where similar applications are gathered. Among these services are terrestrial or space services, fixed, mobile or broadcasting services and so on. Each of them is defined by a limited set of basic features: Chapter 5 will give some views on the main ones. Of course, a band which is allocated to a particular service should be perfectly adapted to these features and to the foreseen needs. A band may be allocated to one exclusive service or to several ones, with priorities. Technical rules are specified to make possible the sharing of a common band by different services. This first regulation layer is internationally negotiated within the ITU and is made compulsory through the Radio Regulations.

At national or regional level, more precise choices and some adaptations of the Regulations are necessary, taking into account *harmonization* concerns after a debate with partner countries, in view of a practical implementation of *applications*. The administrations decide the authorized applications in each band or sub-band.

They can divide it for different applications and plan it into channels. They may specify the standards to be used and complementary technical rules. They may also delegate the management of some bands to particular regulators. All these matters could be called *national allocation* (or utilization) decisions. They are under the responsibility of state governments whose administrations decide on them, taking onboard experts advice and regional harmonization directions.

For better spectrum efficiency to the benefit of a particular service or application, the allocated band can be formatted and planned. It is divided into standard frequency modules, called channels. These channels can be shared between different applications, operators, users, stations on a particular geographic area. The channel width and arrangement depend on the chosen service or application and the system standards. Such a channel organization is called a *plan*. It may be decided by the Radio Regulations, the harmonization bodies or the national administrations.

To facilitate the management of a band between neighboring countries and to minimize cross-border interference problems, a spectrum resource can be sub-divided from the beginning and shared between different geographical areas and countries. This sharing is called *allotment* and is generally part of a plan which indicates which channels are available in the different parts of the area concerned, notably in the different countries, at least along their borders.

Finally, the administration, in every state, or an independent regulator in charge of spectrum management, *assigns* precise bands or channels to operators in the different parts of the national territory, after some possible coordination procedures. Technical and administrative constraints are included in the relevant licenses or authorizations. These may concern detailed electromagnetic compatibilities of systems in the field, but also a number of general policy concerns. Financial arrangements, such as taxes, are also associated.

The ITU constitution (Art 1-2) clearly states the responsibilities. ITU allocates the spectrum to services, makes channel allotments through international plans, may register the national assignments and orbital positions for space systems. All other tasks and procedures are the national administration's responsibility.

As a conclusion, spectrum management, particularly spectrum sharing, combines technical rules and administrative procedures. The technical rules, as issued by engineers, are based on physical laws as described in the first chapters. The administrative procedures are necessary to protect user rights. All of them are a basis on which the legal status of the radio applications is built. Technical and administrative regulations should progress together in order that the spectrum, a limited resource, can always be adapted to the permanent growth of needs.

Chapter 5

Some Regulated Services

Among a great number of different regulatory services, as listed by the Radio Regulations, some are regarded in the technical debate as major categories with exemplary characteristics. This does not mean that any scale of importance may be envisaged between services. Specific services such as radionavigation, for example, are fundamental for ship or airplane safety, but they are dedicated to specific uses and cannot be considered as typical for a great number of applications. Here will be mainly described with some details a limited set of major services whose basic characteristics can be found in many circumstances. For each of them, a technical scheme will be proposed which describes the main conditions for their implementation and the corresponding features of the communication links. A short overview will be given of other services in a last section.

It should be remembered that regulatory services are not practical applications but theoretical categories in which many different actual applications with similar basic technical features can be classified, being managed in the same way by the regulators.

Firstly, some terrestrial services are considered, followed by satellite services. Such a difference is clearly stated by the Regulations which define the "terrestrial services" as different from "space services" which are, practically, services using a satellite platform.

5.1. The fixed service

Apart from the maritime mobile service, the fixed service (or terrestrial fixed service) is the oldest radiocommunication service. It concerns direct links between fixed points on the Earth whose position is well known. Of course, terrestrial intermediate relays are authorized between the two end stations: each leap of the link is then considered a fixed service link. In most conditions the service is duplex (see Chapter 2, section 2.4) which means that two different frequency bands are needed to insure bilateral information communication. However, there are also unilateral fixed services: as an example a link to carry audio or video programs from a studio to a broadcasting station.

The fundamental feature in a fixed service is that the geographical positions of transmitting and receiving radio stations are known with accuracy. Thus, an engineer can extensively use any technical means to confine the electromagnetic field in the appropriate direction. Among these means, antenna directivity is a major tool to increase the communication range, to limit the wave intensity outside the useful link and the incoming noise into the receiver. The antenna directivity properties are used twice, in both transmitting and receiving stations to improve the link efficiency.

Considering the C/N or C/I value (see Chapter 2, section 2.3) as the basic criteria for a link ability, the received carrier level, C, depends directly on transmitting and receiving antenna gains (see Chapter 1, section 1.4). Simultaneously, the noise, N, or the interferences, I, which enter the demodulator, can be minimized if low antenna side lobes are provided in the receiving station. In the same way, low secondary lobes in the transmitting station keep the unwanted waves at a minimum level and do not produce radio noise in the other directions.

In addition considering the fact that transmitting and receiving stations are located on the ground, appropriate situations can be chosen and large antenna structures installed without major difficulties or excessive costs. They can also be directed with great precision. With such conditions, the terrestrial fixed service links profit by the most favorable parameters for the highest radioelectric efficiency. They can use the most efficient means to control the waves. The wave containment is achieved by cautious engineering of every point-to-point link, one after the other.

Of course, the technical means which are used to contain waves, in view of reusing frequencies at some distance, are more or less efficient and complex, depending on the fixed link characteristics, the application and the frequency band.

Many different kinds of terrestrial fixed services are met, depending on needs. Historically, the oldest fixed services have been very long distance links using decametric waves, 3 to 40 MHz, using huge rhombic wire antennae, thanks to the ionosphere reflection. The directivity and efficiency of such systems were low. Frequency bands were too limited to satisfy the demand, even with very restricted modulation bandwidth. The communication quality was poor, due to fast changes in the transmission path. However, at that time, there were no other possibilities for long distance radio communications. Today, these links have nearly disappeared and have been replaced by high quality fixed satellite services.

The most recent terrestrial fixed service applications are line-of-sight short links, typically about 10 kilometers long or less, established between masts or buildings through parabolic antennae with a diameter of several decimeters, using SHF or EHF waves with digital modulations. As an example, for this purpose, CEPT recommends a channel arrangement based on a 3.5 MHz elementary module in the bands 24.5-29.5, 31.8-33.4 GHz (CEPT ERC/REC 01-02). Such fixed links are commonly used to connect distant stations, such as mobile relay stations, to the basic infrastructure networks. They provide an excellent directivity, are easily installed and are only disturbed by rainfall. Such characteristics are compatible with an extensive frequency reuse. The same frequency bands can also be used for point to multi-point networks.

Figure 5.1. *Fixed service. This radio relay is part of a terrestrial fixed service. Two parabolic antennae, back to back, in the 13 GHz band, provide telecommunications links from a fixed station*

Between such extreme applications, many different kinds of fixed radio links exist. They were much used in Europe as terrestrial long distance infrastructures for the public telephone network between 1960 and 1990. At that time, national radio telecommunications backbones were created, from town to town, using mostly centimetric bands (SHF), notably at 6-7 GHz (5,925-7,125 MHz) or between 10.7 and 19.7 GHz. High towers were built in the country to install the relays, with a free-space propagation range of about 50 km. Now, most of these links are no longer used, being replaced by optical fiber cables. These fixed service frequencies are shared with fixed satellite services.

The preferred antennae for the terrestrial fixed service should provide the best directivity, such as parabolic dishes and horns. The most appropriate frequency bands for long distance links are SHF. However, for a few fixed service links still operating in VHF or UHF bands, semi-directive Yagi antennae may be used.

Progressively, the UHF bands have been abandoned by the fixed service for the mobile service benefit and the lowest part of the SHF band (3-4 GHz) is also considered for future mobile services (WRC 2007). More generally, with the exception of customer or remote station connections, the terrestrial fixed service importance is decreasing in countries, such as Europe, where dense cable infrastructures are available. Spectrum is so scarce that it should not be used for links which can be satisfied with inexpensive cables. Furthermore, the terrestrial radio fixed service cannot easily cope with the digital information rate increase needed by modern multimedia applications. On the other hand, in a number of countries where cables are not available so much, the terrestrial radio fixed service remains an efficient and cheap tool to implement large telecommunications networks rapidly.

A noticeable evolution of the fixed service appeared recently with new point-multipoint networks, WiMAX or BWA (broadband wireless access), with different standards such as IEEE 802.16 or ETSI HiperMan. For traditional fixed service applications, with point-to-point links, all stations are equivalent. On the contrary, in WiMAX networks, mainly used for business users, a master station provides a local coverage and connects many customer stations spread over this service area, as soon as they want to be part of the service. Such a digital network makes use of multiple access protocols (see Chapter 2, section 2.4).

A WiMAX network is a terrestrial fixed service. It can use the same frequency bands as the traditional fixed radio links, typically 3.4-3.8 GHz. However, such systems, due to their local structure, cannot confine waves as efficiently as traditional fixed services, or at least the downlink waves from the master station to the customer premises, since they must cover a whole area where new customers will appear later in the future. Thus, it can be thought that WiMAX networks are intermediate structures between fixed and mobile services at a point where recent standard versions (802.16e) allow communication with mobiles.

In fact, many new applications now appear intermediate between fixed and mobile services. This is the case of high density fixed services when the number of fixed links is so high that their individual management becomes nearly impossible. Let us also consider the growing number of LPD (low power devices), or short range devices (SRD), which can be found everywhere due to the low cost of radio components. They are wireless local networks (RLAN), telecommands, telealarms and so on. Their transmit power is limited (typically lower than 100 mW) and they work without a directive antenna. Thus their useful range remains short: a few meters. Sometimes they may appear to be fixed services, particularly inside buildings where they stay attached to a room or equipment. However, they should be considered as mobile services. The reason is that no spectrum regulator or manager knows where they are and has any information about their technical characteristics, being license free systems.

Increasing care should be taken of such SRD which are mass-market products, in order that they do not harm protected services. As their number grows with new applications, they produce an increasing radioelectric noise. As they are freely bought, installed and used, it is nearly impossible to control them. They frequently use bands primarily allocated to some other services but which are also allocated to the "mobile service" with a secondary status. Thus, it is important to make provisions to protect any existing primary service in these bands which may be overcome with so many secondary devices. A great number of bands are progressively opened by national or regional regulators to such SRD applications, throughout the spectrum, with appropriate usage designation, depending on the required characteristics.

It can be thought that the current years are a transition period where many fixed service bands are commonly used by more and more "mobile like" applications for individual customers. Regulation adjustments will be needed to take into account such an evolution.

5.2. Mobile services

Mobile services were born at the very beginning of radio, such as maritime mobile services, to keep large ships connected to the land. The purpose was safety, at first, but also ordinary communications. In both cases, radio was unchallenged by any other means. However, the major expansion of mobile services, as a mass market, is recent, due to the success of the portable mobile phone.

Today a lot of different regulatory services qualify as "mobile":

– terrestrial mobile;

– maritime mobile;

– aeronautical mobile;

with some variants. Corresponding satellite services are also provided.

The mobile service importance is increasing, not only because of the growing number of mobile phones throughout the world. Mobility fits with our changing modern society where people are always going here and there, where customers and workers are traveling, with the need for all to keep in touch with their correspondents or to be linked with various information services, which are more and more sophisticated.

All experts estimate that this impressive growth of mobile communications justifies that much more spectrum be allocated to their needs, since radio is the only suitable medium to connect freely moving terminals.

However, the main character of mobile services, from the Radio Regulations point of view, is not to connect moving stations, nor stations changing their place, but stations which cannot be precisely located in a service area to focus waves towards them. In fact, in most cases, the networks are designed so that communications can be established with moving terminals, but the criterion is that it is not possible to aim at them.

In the other direction, terminals which stay somewhere in a mobile service area, do not know the position of the corresponding radio terminal. It may be a fixed station, relay or base station, or another mobile terminal. For the wave containment objective, the context of mobile services is opposite to the fixed services one.

As a matter of fact, it is wrong to say that the mobile station position is unknown or that mobiles do not know where their correspondents are located. On the contrary, this information is more and more available and accurate, to a point where it may be provided to customers as a commercial service, with a precision of a few meters. The difficulty or impossibility is to use this information to direct waves in such a way that they stay permanently confined on the changing path between the terminals, with efficiency similar to the fixed service.

It can be imagined that radio beams are directed from a fixed station towards any mobile one, depending on its communication needs. This technique may be considered for downlink connections with the help of fixed, electronically steerable antennae which can precisely shape and direct radio waves. It cannot be practically implemented for the uplink since the size and position of the mobile antenna are incompatible with any directivity (see Chapter 1, section 1.3).

Equivalent results can be obtained by digital signal processing, combining received or transmitted signals from different base stations spread over the service area. The calculation is equivalent to a "space filter" which extracts from a global radio noise the only signal coming from a particular spot or, in the other direction, creates a peak signal at this precise spot. However, this "virtual" directivity only exists for the radio stations which are part of the network and have the signal processing "keys". For the other radio stations, outside the network, which may be affected by interferences, the transmitted waves are received as an isotropic noise, looking exactly like a traditional mobile service network spectrum.

Such terrestrial mobile networks using real or virtually steerable wave beams, are still experimental. Considering the existing commercial mobile networks, as they are today, they use non-directive links.

Mobile telephone networks, for example, are built using two different types of radio stations, fixed and mobile:

– the fixed stations, or "base stations", are either isotropic or semi-directive in the horizontal plane (however clearly directive in a vertical plane). They cover a wide service area around or in-front of them;

– the mobile and portable stations are almost isotropic.

Considering the radio link equations (see Chapter 1, section 1.4), the power budget for mobile services looks very different from the fixed services, since the antenna gains are very low, even negative. Furthermore, without antenna directivity, the receivers are more sensitive to interfering waves. Thus, the path attenuation should be very limited to maintain the C/I ratio above a threshold. This is a reason why, for mobile communications, lower frequencies are preferred than for fixed services since the path attenuation increases as the squared frequency on the same distance.

Until the end of the 20th century, most mobile networks used VHF and low UHF bands, about 75 MHz, 150 MHz, 450 MHz. In such bands, it is easy to create with omnidirectional antennae wide coverage areas, several kilometers wide, well suited to low traffic and local professional services. Today, thousands of such networks still exist, dedicated to business or security matters (PMR or private mobile radiotelephone and trunk networks).

The purpose of public mobile telephone networks with a cellular structure is different. They should provide a common telecommunications service to a maximum of customers spread over the widest possible area, using a limited frequency band. Thus, it is necessary to reuse the radio channels as efficiently as possible.

In this context, the transmitting sites should be multiplied, each one covering a limited zone, one cell. The frequency channels used by every cell are planned in such a way that no two contiguous cells use the same channels. Based on a geometric scheme, a contiguous tiling of the country can be carried out using these cells. Such planning is well represented by hexagonal cells placed side by side (see Figure 3.8). In urban areas, a common engineering practice is to install different antennae on a platform, each one covering an angular sector of 120°. With three antennae, the whole service zone around a base station site is covered.

The antennae are carefully designed to be very directive in a vertical plane with a slight "tilt" to the ground in order that the wave beams be concentrated around the base station, not too close, not too far from it. With such conditions, it should be noted that the electromagnetic field intensity is very low at the station foot (an important fact for public health concerns).

Figure 5.2a. *Cellular radiotelephone platform (GSM). The angular division of the coverage should be noted, with three antennae oriented in different directions, 120° from one another. The 900 MHz and 1,800 MHz bands are simultaneously used thanks to two different antenna sets. A slight tilt to the ground optimizes the wave efficiency around the station. There can also be seen, on this platform, two common Yagi antennae for TV reception, one UHF, one VHF*

Figure 5.2b. *Radiation sectors. In the horizontal plane, each GSM antenna covers an angular sector. Here can be seen two sets of antennae, each one with a 60° radiation pattern. Thus, six antennae are needed to cover the whole surrounding area*

Figure 5.2c. *Vertical directivity. It is necessary to cover the whole surrounding area in a horizontal plane. On the contrary, the radio field should be concentrated in a vertical plane. A parallel dipole array provides such a radiation pattern. The grey scale shows that the field intensity decreases rapidly with distance. This intensity is very low at the foot of the transmitting station*

The basic techniques used for the cellular networks do not allow much control over the individual links between the mobiles and the base stations (however the transmit power of mobiles can be adjusted, call by call, depending on their distance to the base station). The wave containment is global within a cell. It can be achieved by appropriate control of the transmitted powers and careful design of the fixed station radiation pattern.

With such engineering basis, various cell sizes can be obtained, from a few kilometers to hundreds of meters in diameter, depending on the required traffic density. Inside a cell, no directivity of individual radio beams can be achieved. The relays and mobiles work together in a radioelectric "bubble" which is designed not to interfere with outside systems. Very large networks can be built by placing cells side by side. When a mobile is moving from one cell to another, it should connect to different base stations, with different frequencies ("handover" procedure), following orders which are decided by the network management software, based on the received field intensity. The terrestrial public mobile networks (in Europe GSM and UMTS networks) are built according to these principles. The same technical features can be used for maritime and aeronautical mobile networks but their cells are much larger, several hundred kilometers wide.

The first cellular public mobile telephone network, AMPS, was created in the USA in about 1980. It used 800-900 MHz frequency bands. The first European system, NMT, was developed a little later in the Scandinavian countries, using bands around 450 MHz. Both of them were analog systems. When the market increased, new bands were necessary to cope with the increasing demand. At the beginning of the 1990s, Europe chose the 900 MHz and 1,800 MHz bands to develop a pan-European digital network, GSM. More precisely, in Europe, the GSM900 uses the 890-915 MHz and 935-960 MHz bands, with a possible extension in the 880-890 MHz and 925-935 MHz bands. The GSM1800 occupies the 1,710-1,785 MHz and 1,805-1,880 MHz bands.

The third generation of radiotelephone systems, called IMT2000 and UMTS in Europe, works at about 2 GHz (1,900-1,980 MHz, 2,010-2,025 MHz, 2,110-2,170 MHz, in Europe). It is foreseen that it will extend to 2,520-2,670 MHz before 2010. New bands are already sought for the fourth generation systems but they can only be found above 3 GHz. Thus, mobile services will have replaced all fixed services in the UHF band. With such high frequencies, cell sizes get smaller and many more cells are necessary to cover a given area than with the first generation equipment. An advantage is that the frequency reuse can be still more intensive with very high spectrum efficiency.

When looking at these ever increasing frequencies, it can be estimated that they are not suited to covering wide low traffic areas, either for governmental or professional applications. Bands around 400 MHz would be more appropriate and the corresponding networks less expensive. A typical need concerns the security forces, such as police, fire brigades, medical care, which deserve modern dedicated integrated networks, covering nationwide areas or even greater ones with adapted standards, TETRA or TETRAPOL as examples. Europe has chosen the 380-385 and 390-395 MHz bands for this purpose.

The mobile service is a domain where international standardization and spectrum harmonization are the most useful, more than the fixed service. An obvious reason is that people travel and wish to use their terminal everywhere to obtain access to familiar applications. This matter is very controversial: should the regulation be technology neutral or is it an appropriate tool to create a unified service, worldwide? Such a debate took place, for example, between Europe and the USA when implementing the third generation mobile system in Europe, UMTS.

It has been said that a "mobile service" is not necessarily mobile and that mobility, as understood by the Radio Regulations, is not directly related to the fact that terminals can communicate when moving. It is clear, however, that such mobility is often a major objective to be reached by the systems. From the point of view of the Regulations, considering the spectrum management, a radiocommunication service is said to be "mobile" when the technical rules and administrative practices of the fixed service, concerning the wave containment and the coordination procedures between radio links, cannot be applied. It can be thought that, in the future, this mobility concept will prevail on the fixed service scheme to satisfy customer needs. However, engineers will try to make the mobile services using recipes from the fixed service to improve the wave containment and spectrum efficiency using real-time adaptive techniques. A regulatory convergence between mobile and many fixed applications could perhaps happen later (see Chapter 14, section 14.1).

5.3. Broadcasting

To broadcast means to distribute common information over a geographical area, in such a way that any receiver within this area can receive the same message. The radio link is unilateral.

This technical definition is not sufficient. The regulation requests that the transmitted message be public and cannot be considered as a private correspondence. Sound or TV programs are broadcast. On the contrary a paging system which delivers individual messages to customers is not considered to be broadcasting, although the radio techniques are nearly the same. This should be considered as a mobile service. However, there is a fine line between such applications and compromises should be found about the bands to be used, depending on the context.

Practically, the major broadcasting services are "radio", such as audio programs, and TV. However, a lot of data broadcasting services have been experienced for a long time: weather and traffic forecasts, financial information, news, etc., sometimes associated with TV signals, but the market response has been poor. For this purpose, broadcasting services are competing with mobile services or the Internet. A recent case for such competition, in Europe, is a touchy debate on the "digital dividend". As TV broadcasting becomes digital, some frequency resources are made available. The question is to decide what they will be used for? Either multimedia broadcasting services or advanced multimedia mobile services? This will be discussed later.

Sound radio was the first mass market of broadcasting services. Its huge influence on public opinion was soon demonstrated, particularly during the period 1920-1940, and governments felt directly concerned by this new medium. This early awareness of politicians for broadcasting had a strong impact on the relevant spectrum management. Among other features, the following ones can be pointed out:

– a clear priority is given to the program contents rather than to the radio medium technical constraints. Cultural and political criteria prevail to assign the spectrum resource;

– the service evolutions are rather slow, because they should all be carefully prepared and managed, not to hurt public habits. The great number of receivers makes the changes of standards difficult and costly;

– most broadcasting services are free. Thus, the market is directed by broadcaster offers rather than customer demands. The customers, being passive, do not participate in service evolutions or investments. Broadcasters are anxious not to disturb their audience.

Such constraints, which have been common during the whole history of broadcasting services, are partly related to the broadcasting spectrum resource scarcity, until now, due to the poor efficiency of analog modulations. The new digital codes and modulations are going to provide many more content opportunities and may open an era of increased competition. It may be thought that, in this context, the broadcasting management practices will not be so different from the telecommunications practices.

However, it can be estimated that, from the beginning, broadcasting services have been rather conservative, from a spectrum management point of view, and will remain as such for some time. In the long term a convergence with other services will happen, probably with mobile services, when new applications, such as mobile

TV, will appear and make use of distributed broadcasting infrastructures, which may look like mobile cells.

A result of a conservative approach is to draw up long-term frequency plans, built to last. Broadcasting services are often framed by international plans approved by ITU conferences or by regional bodies, such as CEPT. Standard channels are strictly formatted. They are assigned or allotted after careful negotiations. Such practices are the price to pay in order not to create interference conflicts in border regions. Typically, the radio and TV spectrum, in Europe and Region 1, was organized in 1961 by the Stockholm plan, completed in 1989 by a conference in Geneva. This plan has just been adapted to terrestrial digital TV by Regional Radiocommunication Conferences which took place in Geneva in 2004 and 2006.

Leaving aside the historical audio broadcasting using amplitude modulation (now converting to digital DRM) of waves in different bands: long (148.5-283.5 kHz), medium (526.5-1,606.5 kHz) and short (parts of 3.4 MHz to 26.1 MHz), with kilometric, hectometric and decametric wavelengths, modern broadcasting services now mainly use metric and decimetric waves. Today most audio broadcasting stations use frequency modulated (FM) carriers whose frequency is about 100 MHz (87.5-108 MHz), a band perfectly suited to covering urban areas or local zones, even with portable receivers without any antenna gain. TV preferably uses a broad UHF band, whose core is 470-862 MHz, which provides similar coverage conditions with the help of receiving antennae of the Yagi type notably, giving a few decibels gain and protecting the receiver from echoes.

Until now, the TV broadcasting modulation was analog (a vestigial sideband amplitude modulation). It is now rapidly converting with the objective of being fully digital, in Europe, before 2012. This change is going to improve the spectrum efficiency considerably and make possible a direct reception by portable sets or mobile terminals, at least in urban areas. As mentioned above, digitization will also make available some bands for services other than common TV broadcasting (digital dividend). At the same time, new bands are opened to terrestrial broadcasting. Digital audio broadcasting is looking for spectrum resources other than the FM band. In Europe, terrestrial digital audio broadcasting (TDAB) is transmitting at about 1,500 MHz (1,452-1,492 MHz) and in a former TV band at about 200 MHz (174-230 MHz).

Figure 5.3. *Today, broadcasting stations which shape the basic infrastructure of national TV networks are often situated on geographical high points which provide the broadest coverage. They are equipped with high power transmitters. Antennae (typically dipole arrays) are placed on platforms or along a mast and oriented towards different angular sectors in order to cover the whole surrounding country. Such an infrastructure makes the customer antennae orient themselves towards their local broadcasting tower and then reinforces their importance for the long term. All changes have to take care of this situation. Audio broadcasting (FM), being received by terminals without any directive antenna, is not concerned by such a constraint. Digital TV will make possible a reception using portable sets, and thus will be less dependent on the traditional TV broadcasting infrastructures. The platforms on large broadcasting towers are commonly used by other services. Here, a number of parabolic antennae for the fixed service take advantage of their privileged position*

Radio broadcasting networks, mainly for TV, are still organized around some high power transmitting stations installed on top of remarkable high sites or towers. Their antennae are engineered to cover the whole surrounding country with a nearly isotropic field (in the horizontal plane), created by several dipole arrays. The objective is to deliver the same message to a maximum number of customers, from such major stations. Secondary transmitters extend the coverage to remote areas which are not covered by their direct broadcasting.

The TV evolution to digital is going to modify these engineering principles which are better adapted to analog TV. It will also change the broadcaster activity and market for a greater flexibility of networks and contents. The competition between professional actors will increase:

– The radio transmitted power will generally be lowered, ten times less in many cases. Thus, the installation constraints will be eased. Several programs will be carried by a single wave (multiplex).

– Receivers will become more and more mobile and portable. In towns, roof antennae will disappear. Broadcasting stations will be installed at lower places and will spread, just like mobile telecommunications relays.

– Programs being distributed on multiplex, the traditional binary association between one TV channel and one program will no longer exist. This technical feature will greatly impact the broadcasting operator business. An intermediate vehicle, the multiplex, will really become the message to broadcast and will be considered as the basic content for frequency assignment. This evolution may be thought similar to the WAPECS prospect (see Chapter 10).

– Synchronous modulations will become common. Digital broadcasting accepts a synchronous retransmitting of radio waves at the same frequency more easily. Today, when a place is out of reach from an analog broadcasting station, a retransmitter can be locally installed to extend the coverage. However, not to suffer interferences, a different frequency should be used. Thus, this practice needs an additional spectrum resource. With a digital modulation, if a precise synchronization is achieved between transmitters, it is possible to locally retransmit at the same frequency. This property can be used by "gap-filler" equipment which is easily available for local installers to improve the coverage in the vicinity.

– All included, the spectrum efficiency will typically be 10 times better. Thanks to powerful picture coding techniques, the number and quality of TV programs being broadcast in a given frequency band will considerably increase. In the past, terrestrial TV was considered as a capacity limited service, offering only a handful of programs. In the future it will provide many more and change the habits of customers who can choose and be more reactive.

5.4. Satellite services

Radiocommunication satellite services were born in 1957 with the Sputnik launch. This first artificial satellite of the Earth only transmitted a radio beacon signal. However, some years later, a significant communication capacity was offered by Telstar which, in 1962, demonstrated the availability of high quality telecommunications links between terrestrial stations through a satellite. Everybody knows the success story which came later.

The Telstar network used bands at around 4 and 6 GHz (4,145-4,195 and 6,365-6,415 MHz) which were allocated to the terrestrial fixed service. Many fixed microwave links operated in these bands. More generally, the Telstar system used many pieces of equipment and technologies derived from such microwave links. This satellite achievement was so successful that, as soon as the next year, 1963, an international ITU conference decided that satellite services should be included in the Radio Regulations as new regulatory services and should have spectrum resources allocated to them.

The Early Bird satellite was launched in 1965. It showed the major value of a geostationary platform for commercial permanent communications. It used the same bands as Telstar, but enlarged. The next satellite families, designed and managed by the international organization INTELSAT and many other national or regional systems, launched during the 1970s consolidated the technical basis of the fixed satellite service and notably the allocated frequency bands 3,700-4,200 MHz for the space to earth path and 5,925-6,425 MHz for earth to space.

During these first years, a number of technical or regulatory choices were experimented with and found appropriate for the future, concerning the spectrum management, such as the conditions for sharing the same bands between the terrestrial fixed service and the fixed satellite service. The efficiency of the geostationary satellite orbit was also stressed and technical rules issued to share this orbit between several different satellite systems. It should be noted here that, for satellite radiocommunications, the choices of frequencies and orbits are strongly related and should be considered together when designing a new satellite system.

Year after year, a wide range of satellite services have appeared. Nearly every terrestrial service was doubled by a satellite version:

– fixed satellite service;

– mobile satellite service;

– broadcasting satellite service;

– radionavigation satellite service;

and so on, with a diversity of applications.

The fixed satellite service using a geostationary platform has been the reference service, from the beginning of space telecommunications. It was pioneered and mainly developed by the international organization INTELSAT, created in 1964, but also by many regional operators on the different continents, such as EUTELSAT, in Europe, and PANAMSAT, in America (INTELSAT and PANAMSAT merged in 2006 to become a new private company, INTELSAT). At first, the main application was to open fixed communication links for telecommunications operators, notably to carry long distance telephone calls or international TV programs. However, a very popular service made its way on the same satellites: direct TV to customers.

For a time, it was thought that dedicated satellites would be necessary for such an application and some ones were launched. However, the technology progress was so fast that satellite broadcasting was found possible as a direct adaptation of the fixed service. This was a first example of the so-called "convergence" between telecommunications and broadcasting services which is now considered an objective of the "information society". Today, a great number of geostationary satellites for the fixed service are mainly or only broadcasting direct TV.

The possibility of using for the fixed satellite service up and down links of the same frequencies as those of the fixed terrestrial microwave links was a key for success since it was not necessary to find particular bands for the space services, which would have been a nearly impossible task for years. Geometry explains easily why this is possible and how to guarantee the mutual compatibility of the two services. The trick is that terrestrial links are horizontal, whereas the satellite links are oriented towards the sky, at least with a minimum elevation angle. This angular difference is sufficient to avoid interferences between the different systems if their antennae are directive enough. Clearly, this is the case for satellite earth stations because they must have an important gain to compensate for the high path attenuation between a satellite and the Earth. Having a great gain, they are very directive at the same time (see Chapter 1, section 1.3). From such objectives and constraints, it appears that sharing the same spectrum between fixed satellite and terrestrial services is quite an obvious choice, at least for SHF and EHF bands where very directive antennae can be easily designed.

Figure 5.4a. *Satellite earth stations. The satellite earth stations for telecommunications mostly work with geostationary satellites which are placed on a particular position along the orbit. The three front stations are aiming at a common orbital position or very near ones. The one behind is directed towards a different position. The largest antenna works in C band (4 and 6 GHz) and its diameter is about 30 meters*

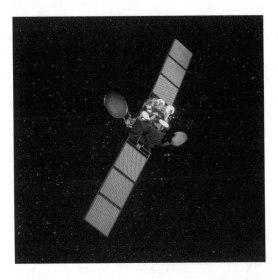

Figure 5.4b. *Telecommunications satellite. Telecommunications satellites are designed to cover a service area on the Earth from their orbital position: either the whole Earth (global beam), or a particular smaller region (spot beam). Here the global beam antennae are placed on the satellite main part. The big dishes for spot beams are deployed on the sides*

Fixed service satellites are relays, in space, which retransmit towards the Earth messages transmitted from the Earth. Thus, two types of links are established: uplinks from the Earth to the satellite, downlinks from the satellite to the Earth. They should be considered separate and different frequency bands are allocated for these two directions.

The wave containment bases, to increase the spectrum efficiency of fixed satellite systems are nearly the same as for terrestrial fixed services, although, at first glance, they may seem very different. Both make the best use of antenna directivity. The antennae of the satellite earth stations (which are the stations, on earth, in relation with a satellite) should have a great gain and directivity. Simultaneously, they must have the lowest side lobes possible either directed to the orbit or, in the horizontal plane, to the terrestrial microwave links (called terrestrial stations).

The ITU has set technical rules concerning the radiation pattern of satellite earth stations, notably their side lobe gain envelope. Their directivity may be so high that they must sometimes "track" the satellite, even a geostationary satellite which moves slightly about its nominal position. This means that the antenna should move automatically to keep aimed at the satellite. As mentioned above, their side lobes are strictly limited. Particular attention is paid to the first side lobe, apart of the main axis, not to interfere with adjacent satellites.

In the same way, the satellite antennae are carefully designed to cover the service area on the Earth exactly. They "illuminate" a definite radiation "print", on the ground, which fits this service area as well as possible. Here too, rules have been issued to limit the radiations outside the main beam. A basic constraint which the satellite should satisfy is to maintain thoroughly its orbital position and "attitude" as well as its antenna direction, in order to keep its radiation print stable, in conformance with its agreed characteristics.

Looking now at satellite mobile services, things are very different: it is necessary to cope with a bad directivity of terminals, notably portable terminals. Thus, it is much more difficult to share common frequencies with terrestrial mobile services since the signal levels to the satellite are nearly equivalent to those of neighboring terrestrial mobile networks. This is the reason why dedicated bands are allocated to satellite mobile services. As an example, for the IMT2000 project of third generation mobile systems, bands 1,980-2,010 and 2,170-2,200 MHz have been allocated to space systems and bands 1,900-1,980, 2,010-2,025 and 2,110-2,170 MHz allocated to terrestrial systems (such as UMTS). However, the links between the satellite and the Earth which carry communications between mobiles and the public fixed telephone network are fixed satellite services and use the corresponding allocated bands.

It is a typical feature of mobile satellite services to need two different duplex frequency sets: a duplex set for the links between mobiles and satellites and a duplex set between the satellite and earth stations which are an interface with the general telecommunications network. The mobile set is typical and dedicated to the satellite mobile service, the "fixed" set is part of the fixed satellite service.

Satellite broadcasting has found its way after a hesitant period. For TV broadcasting, the frequency scheme looks very much as for the fixed service. Programs are transmitted to the satellite via an earth station which uses fixed service uplink carriers. They are broadcast to earth on a beam which covers the appropriate radiation print. The bands are those of the fixed service or ones dedicated to satellite broadcasting at about 11 and 12 GHz. In Europe, the satellite networks EUTELSAT and SES-ASTRA use all the available bands from 10.70 to 12.75 GHz. The customer receiving antennae are fixed satellite dishes whose diameter is about one meter, carefully oriented towards a geostationary satellite cluster installed at a very definite orbital position. Their directivity results from a compromise: getting the maximum gain, limiting the dish size, avoiding any satellite tracking. Some other satellite networks, part of the fixed satellite service, called VSAT, without customer uplinks, are used to transmit data to companies. They look very similar to broadcasting systems.

It can be mentioned that as the number of TV programs is increasing, they may be broadcast by different satellites placed on very close orbital positions. A customer antenna pointed to such a cluster simultaneously receives carriers transmitted from the different satellites which share between them the available frequency bands. This is typically what ASTRA and EUTELSAT are doing over Europe: each operator places different satellites on its reference orbital position. Another advantage is to provide a self-redundancy of the satellite cluster.

For satellite audio broadcasting, still different technical choices should be made. The receivers, on the ground, are commonly fitted out with slightly directive and poor gain antennae, looking like mobile terminals. Thus, the frequency bands should be dedicated and much lower than for TV satellite broadcasting, at about 1,500 MHz, for example.

5.5. Geo and non-geo systems

The satellite services are different whether they are provided by geo or non-geosatellites and the corresponding frequency management differs too. Non-geosatellites have been used for a long time: the first telecommunications satellite, TELSTAR, and the Russian family, MOLNYA, were non-geo platforms, placed at various orbits adapted to their purpose. However, these rather simple systems with limited earth coverage have been almost ignored over the years when geostationary satellites became the reference technology. At the end of the 1990s, non-geostationary civil telecommunications satellites were found attractive again. Futuristic projects were promoted by private operators, such as IRIDIUM, GLOBALSTAR, ICO, TELEDESIC, SKYBRIDGE and others. Using satellite constellations, with tens or even hundreds of space platforms, either running on low orbits (LEO, about 1,000 km) or medium orbits (MEO, about 10,000 km), these networks were designed to offer a real worldwide coverage, even on polar areas, with services more sophisticated than geostationary systems: their marketing objective was to provide direct access to the satellite for any customer on Earth. However, these great projects which concerned fixed or mobile applications, did not have commercial success, being too costly and less easy to use than terrestrial networks. They collapsed during the telecommunications financial crisis in about 2000.

The last word has not been written about non-geo systems. Today they are used for a great number of applications, outside the civil telecommunications domain, such as military purposes, radionavigation, environmental studies, etc. They may come back before many years have passed for particular communication objectives, as a complement to terrestrial services.

Due to their movement in the sky, a permanent communication with non-geosatellites needs terminals which can work in nearly any direction. Thus the customer equipment may be fitted out with an omnidirectional antenna or with a steerable directive antenna which tracks the satellites. Such a feature makes the frequency coordination more difficult with the terrestrial fixed services or geostationary satellite systems which use the same frequency band. The coordination appears still more complex between different non-geosatellite networks. This was a reason why it was feared, at the beginning, that an operator who implements such an important non-geosatellite constellation, would use the whole available spectrum resource only for its own benefit and gain a monopoly. As an example, during the World Radio Conference of Geneva 1995, the American administration requested that a duplex band, 500 MHz broad, about 20 and 30 GHz, be made available for a non-geosatellite project, TELEDESIC, in such a way as, practically, it would be a monopoly.

However, the next world conferences made progress to invent technical methods, called mitigation, which make it easier to share the frequency bands between geo, non-geo and terrestrial systems, at least for the fixed services. Their principle is to interrupt a non-geo communication link when it crosses the equatorial plane where the geostationary orbit lays or when it becomes earth tangential. Such link management constraints make the network software more complex for non-geo systems but they greatly improve the spectrum efficiency. The acceptance by the ITU of such a sophisticated control method made possible a global sharing of all bands allocated to the fixed service by geo, non-geo and terrestrial systems, having extended the traditional wave containment rules to the three dimensions of space.

Some other types of telecommunications services are used by satellite systems for themselves. They are more technical, less well known, but necessary, such as links between satellites (inter-satellite service), or control and command links to manage satellites from the Earth (space operation service). It is not useful to give any more indication about them but the Regulations allocate appropriate bands for such purposes and make them compatible with other services.

5.6. Some other regulatory services

Communication is, by far, the main purpose of radio systems. However, they can be used for many other objectives.

Firstly, transmitting radio waves is not only a human activity; it is a natural phenomenon which was studied as early as the 19th century (Edison, Kennedy, 1890). Various radioelectric signals are emitted by a number of celestial bodies. They bring us much information about them and the space medium that they have traveled across. Studying these natural signals is the task of radioastronomy whose scientific importance has grown over the past century. Some bands are protected in order that they can be observed without any disturbance. In these bands, any human emission can be forbidden. Most of these bands are above 10 GHz but some important ones are UHF, such as 1,400-1,427 MHz (frequency 1,420 MHz of neutral hydrogen). Some bands are not forbidden but protected against terrestrial emissions, typically with a radio silence area around radiotelescopes (various bands between 1,612 and 1,720 MHz of the OH radical).

Radioastronomy protection becomes more and more stringent and exacting. It may conflict with other telecommunications services due to their growing activities. Today some interferometer radiotelescopes extend to several kilometers but, in the future, space interferometers with 1,000 km spans to provide extremely accurate angular measures can be imagined. The receiver sensitivity also increases and looks for the best radio silence possible (sensitivity better than (10) exp-30 $W/m^2/Hz$ obtained by signal processing on long observation periods). Moreover, for deep space observation, radioastronomy needs wider frequency bands to cope with the red shift of natural sources at the cosmos border. As an example, the observation of hydrogen or hydroxyl rays may require listening to signals from 1,200 MHz to 1,725 MHz. Such demands cannot be easily satisfied, notably with the development of satellite systems which may cross the receive pattern of telescopes or with the increasing number of mobile terminals which cannot be strictly banned from their vicinity.

Radio stations are also used to measure chemical contents of the terrestrial atmosphere such as ozone (110 and 142 MHz) or water vapor (22 GHz). The observation of Earth by satellites uses similar techniques with signals about 21 or 37 GHz. The World Radio Conference, Geneva 2000, has accepted that any band above 275 GHz which is not allocated, can be freely used for research. Some scientists are already studying spectral rays related to the interstellar matter with frequencies as high as 400 GHz. Many applications which are among the most important for transportation security and defense are called radiodetermination services, with such categories as radiolocation or radionavigation services.

Figure 5.5. *Radiolocation: aeronautical radar*

Radiolocation is a technique to measure, by means of radio waves, the position, direction and speed of distant objects. Among the corresponding equipment, radars are the most well known. They use very different frequency bands: from large airport radars for the aeronautical domain survey which are working about 1.3 GHz, to the very small ones for car speed control or local movement detection using much higher bands (24 GHz). Radars were invented a few years before the Second World War. They proved to be very efficient during the conflict to detect planes at a very long distance. Despite being frequency limited, at the beginning, due to the available electronic tubes (about 200 MHz), and thus poorly directive, they made decisive progress when they could use microwaves with the magnetron invention. Today, thanks to cheap solid-state components, many types of radars are available for short distance applications in millimetric bands. However, for such local needs, equipment may also be considered as relevant to the mobile service (LPD).

In fact, some radars are radio probes. They transmit a calibrated signal in a direction to get information about the transmission path from the received signal or from an echo. Being installed on the ground, or on a satellite or in a plane they are scientific instruments for environmental studies. As an example, meteorology commonly uses them, notably for hydrological research.

A typical feature for many radiolocation services is that they use pulsed carriers with a very broad spectrum and a huge instantaneous power. Considering too that their received signal is weak, near the noise level, and that an objective is to detect a significant message in such extreme conditions, it can be understood that one of the main problems for their spectrum management is to insure an electromagnetic compatibility with other systems. This compatibility concerns systems in the same band, of course, but also those which occupy neighboring bands, due to the importance of out-of-band signals which may be created by pulsed equipment.

Radionavigation is a service which helps mobiles to determine, by themselves, their geographical position with the assistance of radio beacons installed on the ground or on satellites. Radionavigation is provided for terrestrial, maritime or aeronautical mobiles and the Regulations allocate bands for these different contexts. Among the radionavigation applications, the most popular, today, are provided by dedicated satellite systems such as GPS or the European project GALILEO which use bands between 1,160 and 1,610 MHz. Their basic principle is to measure the distance between the mobile and radio beacons on board satellites whose position is known with the greatest accuracy, from the transit time of waves. The location precision depends directly on the time stability of the master clocks.

It can be noted here that goniometry (or direction finding) is not a regulatory service. It aims to measure the direction of a distant radio transmitter. This is a purely passive activity which does not need any particular radio resource.

To be exhaustive, a great number of other dedicated services should be mentioned, being more or less common, but a list would be boring. To illustrate this variety, let us mention standard frequency and time signal services which transmit all around the world calibrated waves as a reference for the international time. Let us also mention ISM applications (ISM as industrial, scientific and medical equipment, such as microwave ovens). These ISM make use of waves in a closed place for purposes other than communication. They should not be mistaken with LPD (low power devices) which are real telecommunications devices. In any case, all services which show a real need of radio resources are considered by the ITU and bands allocated.

As a conclusion, let us pay attention to amateurs. From the beginning, their radio communications have been considered as a true regulatory service. The international bodies consider them to be a valuable educative tool and a possible operational help in some rescue circumstances. Radio amateur activities are organized by strict international rules. This is why they are allocated frequency bands, exclusive or shared ones, in most spectrum parts, from 1.8 MHz to 250 GHz. Amateurs are concerned with satellite and terrestrial services as well.

Chapter 6

Recent Evolutions of Radio Services

We should now consider the actual development of radio services over recent years to understand their tendencies, needs and constraints, and to give a general idea of their past evolution.

Up to now, we have examined the basic conditions, technical, legal or administrative, which frame radiocommunication management and make it consistent and stable, such as the laws of physics, information theory, geography, international legislation, regulatory services and others. However, such constraints are mainly static and cannot explain the movements that spectrum regulators must permanently direct and manage. It is necessary to consider the services, as seen from the customer and operator points of view, to take appropriate decisions on activities such as:

– allocation and planning of frequency bands with pertinent technical dispositions;

– rule setting for coordination and compatibility;

– resource assignment to users;

– spectrum monitoring and procedure enforcement.

The international Radio Regulations are a relatively abstract document. They remain unchanged between World Radio Conferences, whatever happens. They try to be as neutral as possible about the different actual applications which are classified as the same regulatory service. Officially, they do not take into consideration social characteristics: business or governmental for example, nor the countries where they are promoted in the same Region. They make no distinction between private, public, corporate or administrative networks. They maintain a technology and standard independent approach, with no direct reference to particular projects or political, economical and societal objectives. However, this neutrality is partly formal: practical concerns and objectives are the strongest reasons to decide on future regulatory adaptations but they will not appear as such in the Regulations.

On the contrary, the different national administrations are responsible for the actual radio applications available in their country, on short and long terms. They should comply with the Regulations but must satisfy the day to day demand of operators and users which they rule, with a view on national objectives and priorities, in conformance with the local legislation.

To understand spectrum management evolutions, it is often necessary to look at the strategy of major countries and regions, concerning their radio resources, and their commitment, as administrations, to support or oppose any project. This policy of every administration explains its attitude during international debates. The Regulations, as a "neutral" and abstract document, are in fact a negotiated agreement between countries. It is a mirror of accepted compromises between initial national points of view and strategies concerning precise and realistic projects.

At a later stage, and at a national level, as explained in Chapter 4, the Regulations are only seen as a canvas, a general framework which has to be filled and completed with detailed decisions and specifications, in order that the spectrum resource can be assigned to actual users for effective applications. Here and now it is clear that these applications are not theoretical categories but real implementations which should satisfy users. The success or failure of practical investments depend on the corresponding national decisions, with technical, financial or administrative parameters. The national administrations, more than the ITU, take on important responsibilities when they decide the authorized services and systems, choose operators, enforce local constraints and regulations.

Such national spectrum policies must be clearly explained and published to be efficient. As the first resource manager, the administration of every country has to inform the citizens on the spectrum availability for different applications, their rights and duties as users for frequency assignment. It should apply transparent procedures to enforce and monitor the rules. The policy has to be described in publicly available documents based on local laws and general regulations. Open databases on radio systems and networks should be developed. The impact of radio on our modern societies has become so important that dedicated public experts and independent regulators are useful for such a public transparency on the spectrum matters.

The growing importance of rational common frequency management for the best efficiency of radio services in the vicinity of national border lines and the customer demand for an open world where the same services are everywhere available, progressively oblige the national administrations to cooperate and harmonize their choices. It may be only a voluntary technical cooperation, guided by informal international expert debates and recommendations, but the harmonization can also be backed and enforced by political bodies, such as the European Union, which are in charge to create a common and unified market between countries. They are becoming more and more concerned with spectrum management, as a powerful tool to achieve common economical objectives in the radiocommunication domain, going further than the technical harmonization of systems. Such common political visions must be studied and formulated in cooperative forums before any coming debate at a political level or within the ITU. They will gain strength and support if they appear to be backed by many participants such as operators, industry and customer representatives, etc. and thus can be accepted as common objectives by administrations, which should be implemented in the same way, with similar and parallel procedures.

6.1. A family snapshot

Is it possible to measure, even approximately, how much of the spectrum is used at a time? It could be very useful but it seems impossible to get an agreement on a common scale.

A megahertz does not have the same use and value in a VHF or EHF band. It cannot be easily compared when it is available in a big city or in a desert. Is an allocated band with primary status equivalent to the same one with secondary status? How can we estimate the relative efficiency of a radiotelephone band whose frequencies are massively reused by small cells and a radioastronomy band where no emission is allowed? How do we appreciate shared bands? Many other questions can be asked and this makes the problem impossible to formulate clearly.

If any measure can be argued, let us however express a "feeling" about the relative weight of the different social uses of the spectrum in our modern societies:

– governmental (defense, police, public security) 40%;

– public commercial uses (civil telecommunications, broadcasting) 40%;

– space (all services included) 10%;

– transportation (aeronautical, maritime) 8%;

– science (radioastronomy, meteorology, etc.) 2%.

These relative weights are considered for frequencies lower than 60 GHz. Governmental and public commercial uses get bands equally distributed throughout the spectrum, from HF to EHF. Transportation is more concerned with HF to UHF bands. Space and science are mainly interested in SHF and EHF.

As spectrum allocation and assignment are controversial matters where politics and public opinion play a significant role, let us summarize the debate for spectrum resources between only three main parties:

– public administrations 40%;

– consumers and operators 40%;

– professional users 20%.

Of course, these approximate figures do not mean at all that the spectrum is strictly shared accordingly. Let us stress again that the Regulations make no such distinctions between social categories. Many bands are simultaneously used by these different parties. However, they may be found useful to get an idea on the relative influence of such major "lobbies".

Within a category, many secondary distinctions take place, corresponding to different branches: various ministries, industries or historical techniques. They act sometimes as partners, sometimes as competitors for the spectrum resource, being backed by dedicated regulators and expert forums. Such traditional professional categories counterbalance the technical opportunities to create convergences on a common use of bands.

As an example, within the "consumers and operators" category, major differences still appear between telecommunications and broadcasting people, due to legislations, markets and other features which oppose the technological trend towards integrated consumer services. Broadcasters control most bands below 1 GHz. Telecommunications operators and free use systems are progressively gaining a major influence above this limit: they cannot easily imagine a common future together, even for hybrid services such as mobile TV or data broadcasting. Similar competitions may take place between governmental uses, for example between defense and police forces.

Hopefully, technologies, notably space technologies, are rapidly progressing and propose modern products which overcome historical borders. Telecommunications satellites for the fixed service have been found to be perfectly adapted to direct TV broadcasting. Radionavigation by GPS or GALILEO satellites fits with all travel means purposes: planes, ships, cars and pedestrians. On the ground, the technical "scheme" WAPECS which would locally distribute high rate digital frames to customers in a definite area appears to be a reference for convergence. Customer fixed and mobile services are progressively merging. Even different "social categories" can be interested in the same products, based on "dual" technologies.

GPS is simultaneously a military system and a consumer market and GSM is now commonly used by many public authorities in place of their traditional private radio networks. Thus, the future will clearly be different from the present situation: users, services, applications will change and spectrum management will follow, accordingly.

However, for the time being, some major application categories are still references for the spectrum management. The corresponding resources are permanently reshaped and distributed, always depending on new needs. The World Radio Conferences, regional harmonization decisions and national frequency assignments modify their structure and use. Let us look at some recent evolutions which have been much discussed within international bodies and which directly interest the consumer mass market.

6.2. Enthusiastic telecommunications

Beyond all question, telecommunications started the main spectrum changes over the last 20 years with an astonishing dynamism, although, in the end, many projects cannot find a real market. The stock exchange crash, in 2000-2001, which severely hit the share value of high technology companies, has only slowed the movement and made promoters more cautious.

Two powerful telecommunications engines pull the information society: the Internet and the mobile phone. Both of them have triggered a radio renewal. Apart from satellite projects which we shall introduce later, radio services for terrestrial telecommunications have experienced a revolution. However, it is without doubt that the most important change for spectrum use was created by the portable mobile phone, notably by the extraordinary success of the European GSM. Of course the radiotelephone has existed for a long time, notably with cellular networks such as the American AMPS or the Nordic NMT which could already address a mass market, but the mobile revolution really began with GSM, in about 1995.

During the World Radiocommunications Conference, Geneva 1979, it was decided to provide mobile services with significant bands. So far, such services were not generally considered strategic, at least for civil applications. Now, a part of the 900 MHz band which was allocated to the fixed services was opened to mobiles. On this basis, CEPT (European Conference for Post and Telecommunications) decided, in 1982, to keep available the bands 890-915 and 935-960 MHz for a pan-European mobile service. Simultaneously, a working group, called GSM, began to specify a new digital mobile system. Five years later, the technical standard was completed and thirteen European countries approved it, in view of a common operational project which was backed by the European Commission. The first two relevant networks were implemented in France and Germany, as soon as 1992. Since that time, the new system has had remarkable success and can be considered a worldwide reference. More than 500 million customers were gained in 2001, with 250 million in Europe. Now, in many countries, the number of mobile phones notably exceeds fixed telephone lines and even the total population. Considering the opinion of the GSM promoters, as expressed in 1992, that a possible market of 20 million terminals could be addressed in 2000, it can be judged that the GSM mobile phone was really a revolution.

This market explosion was due to many specific reasons, apart from enthusiastic customer acceptance for a new tiny personal terminal:

– the digital technology progression and the availability of powerful integrated microcircuits;

– the European unification with its vision of great common achievements;

– the economical and technical competition which followed the end of historical telecommunications monopolies;

– the growing mobility of people.

Such a positive context boosted energy and ambitions to fully reshape the telecommunications business. This was not only a European phenomenon: everywhere in the world, at the same time, a tremendous growing demand was observed for mobiles, but the European strategy to answer this demand was really the best one.

Spectrum organization was adapted to cope with the needs and the growing demand of operators to get new bands assigned. A complementary band, around 1,800 MHz (1,710-1,785 and 1,805-1,880 MHz) was reserved, in Europe, as soon as 1995, to extend the GSM networks. The original band was also increased (880-890 and 925-935 MHz).

In the ITU, experts soon imagined that, at the beginning of the 2000s, a third generation of mobile systems would appear with extended information capacities, which was called FPMLTS (future public mobile land telephone system) and later IMT2000. Many different standards were designed, adapted to this future project, among which the European standard UMTS. New bands were allocated by the World Radio Conference of 1992, on a worldwide basis, to be used by the corresponding networks (1,885-2,025 and 2,110-2,200 MHz). Also, the prospective of a complementary band, for extension, between 2,520 and 2,670 MHz, was adopted in Europe (WRC 2000). Now, in 2007, the debate is open for a fourth generation system and the corresponding bands. There will be some directions given about them at the WRC 2007.

GSM was not the only standard for digital mobile phones around 2000. Many others, Japanese and American, notably, were available during this period but none was so widely used throughout the world. This European product can be considered the reference to illustrate the surge of a new consumer attitude for personal telecommunications.

This success was the origin of other projects to provide mobile personal communications in all circumstances and contexts, for many needs. Radio appeared as the obvious medium to connect customers to the networks, anywhere and anytime, making cable links to their premises and terminals an obsolete constraint. At the same time, the large telecommunications infrastructures got rid of long distance microwave links to adopt optical fibers which, alone, can provide the necessary information transfer capacity required for the information society. It is a clever trade-off. Progressively, the spectrum under 3 GHz is being abandoned by the terrestrial fixed services and made available for mobile services.

In the future, the mobile phone will be the basic personal tool for customers to communicate and interact, while moving, as the personal computer is the heart of private or business information systems. It will compete with many terminals, such as broadcasting receivers or transaction devices. It has already replaced some other limited mobile services, such as paging. The pager, a simple and economical terminal to receive personal unilateral messages, met with great success during the 1980s. It was considered as an ideal substitute to the mobile phone for many customers, being simpler and cheaper, at a time when the mobile phone was seen as a luxury. This idea was so common that a harmonized paging standard was thought necessary in Europe (ERMES, 169 MHz), but it was never widely adopted.

The story is about the same for private mobile radiotelephones (PMR) whose number had been steadily increasing for years to the benefit of company agents or independent users. Such radio networks gave them the freedom to communicate directly with their office within a range of a few kilometers. However, their success is now declining although modern digital products are available. Many independent networks are closed by users who prefer to adopt the public radiotelephone. Furthermore, these public networks adapt themselves to provide special functions which make them comfortable for such business users. As an example, the "push to talk" call procedure can be provided to make calls similar to the familiar simplex practice, the same for a "fleet call", when a same call is simultaneously addressed to a number of mobiles.

Progressively only big companies, trunk network operators or governmental services can afford to implement modern independent radio networks with their own infrastructure. Adapted standards have been developed for them such as TETRA and TETRAPOL in Europe. Public security is probably the main concern for such networks: being frequency independent, they cannot be congested like a public network during an emergency situation. Their messages can be more easily enciphered and the calls managed and routed according to specific rules. In the same way, GSM variants have been designed to cope with particular professional contexts. For example, the European Railway Union has adopted its own system, derived from the GSM, to keep trains connected to the fixed infrastructures.

The market does not answer positively to any technical offer, even when it corresponds to an apparent need. An example is the Terrestrial Flight Telecommunications System (TFTS) which has been planned in Europe, for some time, using the 1,670-1,675 and 1,800-1,805 MHz bands. This project intended to provide commercial telephone calls for plane passengers. A decision was obtained in 1992 when appropriate technologies were available. However, this service which should have been fully deployed about 2005 was not successful and was stopped after 10 years.

Another major international program which failed to achieve its objectives was the IRIDIUM project for personal mobile communications through satellites. The concept was widely promoted, on a worldwide basis, for a public mobile phone network accessible anywhere on Earth, thanks to 66 low orbit satellites. This beautiful idea seemed to be the right one to make telecommunications available to anybody, even where no terrestrial infrastructure existed, which was a common situation in many developing countries at that time. However, the project had to compete with the unexpected dynamism of the terrestrial mobile networks, such as GSM. It was too expensive and could not sustain the comparison with the terrestrial service and terminals for their comfort, price and quality. IRIDIUM, whose commercial launch was predicted in 1998, did not find its way.

A more political reason should perhaps be proposed to explain this unsuccessful story, which illustrates perfectly the difficult debate on harmonization versus technology neutrality. The USA administration opposed the choice of any unique mobile telephone standard by telecommunications regulators, arguing that only free competition between standards was acceptable. According to this open market approach, the European view of a pan-European reference standard, GSM, was inappropriate. From the US point of view, the preferable situation would be that any operator makes its own choice. However, in this context, the need may exist for homogenous world coverage to complement the local particular cellular networks, in order that travelers may communicate from anywhere with a single terminal. A worldwide satellite system seemed to be the right tool and could have given its promoter a major advantage. However, this strategy failed and the international markets preferred to make voluntary harmonized choices rather than give a monopoly to any satellite super network.

As mentioned in Chapter 5 (see section 5.1), by many aspects as seen from the spectrum management point of view, the "last mile" link between the customer premises and the telecommunications networks, if carried by radio, looks like a mobile service rather than a fixed one, being considered by the Radio Regulations as a "high density fixed service" (HDFS) with mixed features between fixed and mobile services. The equipment to build such a connection can use both technologies. As an example, around 2000, many mobile satellite systems which were being studied, were also adapted to fixed customer connections, notably in low density areas and developing countries. Their promoters explained that with minor adjustments to the antenna system or electrical supply, at the customer premises, the satellite networks would fit perfectly with the constraints of the rural telephony and Internet access. In many forums of the ITU development sector, IRIDIUM, GLOBALSTAR and other teams stressed this opportunity of the mobile satellite service.

At the same time, the fixed satellite service was found to be appropriate by some other operators to connect directly most customer premises to the Internet via small individual earth stations. The major projects made use of low orbit satellite constellations, called TELEDESIC or SKYBRIDGE, which were the subject of exciting debates in international meetings. Notably, the huge American project TELEDESIC, with 840 satellites, created a tremendous surprise during the World Radio Conference, Geneva 1995, when requesting two 400 MHz widebands, 18.9-19.3 and 28.7-29.1 GHz. However, this network was never born.

Many less ambitious projects of terrestrial radio networks, point to multipoint, were elaborated to connect locally customer premises or terminals to the global IP (Internet protocol) network. They are notably adapted to the benefit of new telecommunications operators who want to compete with the historical telephone companies and prefer not to depend on the traditional copper customer loops. As a first step, about 2000, much hope was expressed about wireless local loops (WLL) which could use bands allocated to the fixed service around 3.5 GHz and between 27.5 and 29.5 GHz. However, after an enthusiastic period, the market response was disappointing. A new concept was born in about 2003, called WiMAX (worldwide interoperability for microwave access), based on American standards IEEE 802.16, which seems more successful and may offer fixed and mobile services together (see Chapter 5, section 5.1).

Many other original ideas have been explored during the same period to satisfy the new need for wireless high speed data connections between personal computers and IP networks. The RLAN concept (radio local area network) appeared as a simple solution to create low cost, license free access infrastructures, considered, from a regulatory point of view, as a short range device (SRD), which can be installed anywhere. Most professional buildings and public places have been equipped with such local radio networks for the convenience of any user, in their vicinity. Appropriate standards such as IEEE 802.11b (WiFi) are widely available for medium speed data streams, up to a theoretical value of 11 Mbit/s, (2.4 GHz) and IEEE 802.11a or ETSI Hiperlan, for higher rates, up to 54 Mbit/s, theoretical value, (5.2 GHz). Such equipment has met with great success and is praised for its efficiency and instantaneous availability, in spite of some drawbacks concerning the information confidentiality or protection against interference. It can be estimated that it will be kept restricted to purely local connections and cannot be used as a basis for wider area networks.

Within the ITU, during the 1997 and 2000 world conferences, the opportunity for high altitude platforms (HAPS) for high density fixed services (HDFS) was also discussed. They could be installed on new high points such as stratospheric airships and may use very high frequency carriers in EHF bands. Only the future will say whether such ideas are realistic and have any merit when compared to satellite or traditional terrestrial networks.

Finally, let us stress again the extraordinary development of low power devices (LPD) or short range devices (SRD) which are used in all circumstances for countless needs: telecommands, telealarms, HF microphones, portable cameras, wireless connections, etc. Many narrow bands, throughout the spectrum, are allocated to the corresponding mobile service, mainly with a secondary status and without protection against interferences. Some standards have been studied for particular equipment such as Bluetooth to connect together peripherals of multimedia information systems. A CEPT recommendation (ECC/DEC/(04)02) describes some bands available in Europe for SRD with associated technical specifications.

To summarize this extraordinary fountain of new ideas and products for telecommunications via radio, it should be considered that they are all dedicated to the end customer connection, to make access to services easier. Radio is becoming the natural medium to maintain a permanent link between a mobile citizen and the outside world. A moving consumer or worker does not want to be dependent of any fixed wire or engaged by any inflexible contract with an exclusive operator. He does not want to stop and plug in anywhere. He wants only one thing: to be free. Only radio provides such a freedom.

6.3. Hesitant broadcasters

It has already been mentioned that audio broadcasting was the first mass market in the history of radiocommunication services. In fact, the first wireless applications were long distance communications for fixed or maritime services but audio broadcasting gave birth to the "information society", in about 1920. Then TV became the king of the media after the Second World War. Both services together shaped the electronic media market until the arrival of the Internet.

The traditional broadcasting technologies were analog until recent years when direct satellite TV opened the way to digital techniques. However, important progress has been made during the last 50 years: we only need to mention FM audio broadcasting (87.5-108 MHz) which gave us the opportunity to create many local programs with an excellent sound quality. For the last 15 years, satellites have also multiplied the TV programs available to the customer, wherever he is. However, notwithstanding such positive evolutions, it can be estimated that broadcasting services have been rather reluctant to keep pace with the technological trends of other electronic media and the quick changing telecommunications world. They preferred to stay apart with their own technology standards, at least for accessing the customer which is at the very heart of the broadcasting business.

In fact, the digital modulation of the broadcasting carriers is the key to entering the multimedia world. Thus, audio and TV radio broadcasting can merge and compete with the new actors: satellite, cable, mobile, Internet. For the last few years, progressively, every analog service has wondered whether it should go digital to improve its quality and increase its capacity. The most recent idea is DRM (digital radio mondiale), backed by an international group of industrial partners since 1998, with ITU support. It may completely renew the audio broadcasting services in the historical bands, from "long waves" to "short waves", whose carriers are still using traditional amplitude modulations. DRM affords a quality which compares with FM. Thus, in the long term, even the FM band could be interested by such a new modulation. One advantage of DRM is keeping the frequency plan unchanged. It respects the traditional association between one frequency channel and one program and thus does not disturb the spectrum management principles in bands which are used internationally.

At the same time, around 2000, when the idea to change to digital modulations was gaining support, there was some hesitation about the appropriate strategy.

Concerning local audio broadcasting (FM), no real demand and no clear evolution appear. Reasons and arguments to move are not convincing and technical solutions are still debated. Broadcasting data in parallel with audio programs, for example, is compatible with FM and does not justify major changes and investments. Until now, it has been thought preferable to leave the FM band broadcasting infrastructures and programs as they are and to introduce digital programs, with the TDAB standard, in some other bands allocated to this new application (1,452-1,492 MHz and 174-230 MHz, in Europe). However, the market remains doubtful about the future of DAB and other technical methods may be explored. Competition is growing with the Internet to access audio programs and it seems that mobile phones can also play a role as a convenient terminal for this network. Possibly too, a common digital broadcasting infrastructure could be used for TV and audio, using DVB standards. As the DVBT standard is now accepted for terrestrial digital TV and TV broadcasting infrastructures are being massively converted to this new system, is it really necessary to maintain a different network for audio broadcasting?

This question is asked in a wider sense through the WAPECS prospect of the European Union (Wireless Access Policy for Electronic Communication Services), looking for multi-usage radio platforms which would provide many different services to customers.

In spite of such uncertainties, the interest remains high for the audio broadcasting services and their future. Some European countries, such as the UK, are experiencing fair success for TDAB and the digital satellite audio broadcasting (SDAB) is operational. The market will choose between so many opportunities.

Things are now clearer for TV terrestrial broadcasting. It is already adapting to a digital modulation and the UHF spectrum is being reframed in view of this objective. Various strategies are planned in different countries, depending on the relative importance of TV terrestrial broadcasting when compared to cable and satellite distribution. In countries where cable is prevailing, such as Germany, the switch to digital broadcasting can be very quick and on a large scale. On the other hand, in France and other countries where terrestrial broadcasting concerns nearly 80% of the population, a slower pace is necessary with a long period for simulcast of analog and digital programs. A complete transition to digital can be forecast in Europe before 2012 and all European countries are facing this deadline. Two regional broadcasting conferences have taken place, Geneva 2004 and 2006, to adapt the UHF frequency plan (470-862 MHz) as designed by the Stockholm 1961 and Geneva 1989 conferences, to the new digital service in Region 1. Among the benefits for customers, there will be a greater number of programs, an improved picture quality and the possibility to use portable receivers, without an external antenna, in most urban areas. The opportunity for high definition programs and mobile TV reception is also being studied and experimented.

This main technical direction is now adopted and TV broadcasting will soon convert to DVB digital frame broadcasting. But what about the content of the digital streams provided by DVB transmitting stations? How should they preferably be used? What balance should be made between ordinary, HD, mobile programs and other services? Which picture coding standard will be preferred (MPEG2, MPEG4, etc.)? Such debates with their technical and commercial aspects have been constant, throughout the history of TV. Even during the analog TV era, there were disagreements between the promoters of standard or improved quality pictures. Now, digital TV, with so many opportunities, is going to reopen the discussions.

Another question concerns the possible new broadcasting services which justify keeping the allocated broadcasting bands as they are today. Is there really a market for HD TV, mobile TV, new programs and high speed data? Why not limit broadcasting services for their present situation and use the frequency bands which are made available by the digitization for other purposes than more programs? What about new interactive services which would combine telecommunications and broadcasting services, giving more sense to the convergence spirit? This debate between experts is commonly called "digital dividend". Let us hope that it will end with directions which can stand for a long time and take into account the high economical value of the spectrum concerned. As already explained, broadcasting services are long term, several decades, in order that customers are confident and buy terminals. They cannot be convinced by changing objectives.

Today, audio and video messages are brought to the customers via many different media: satellite, cable, terrestrial broadcasting, ADSL, etc. Does terrestrial broadcasting offer real advantages which justify allocation of an important part of the spectrum? In the past, until the end of the 20th century, the messages were distinct, depending on the medium: interactive voice and data communications for the telecommunications network, audio and video for the broadcasting companies. Thus, any question on the broadcasting future was avoided. This is not true any more and the convergence of messages and networks will increase, as stressed by the European Union Commission by 1997. In the future, messages and networks will become independent and competition will increase between networks to carry any message.

Thus, what are the ultimate and strong reasons to maintain a large terrestrial broadcasting service? Many cultural, economic, technical or commercial aspects should be considered. However, as for telecommunications, it can be thought that the best argument for radio is the customer's personal freedom. Portability, mobility, independence from any exclusive link or contract are attractive features which terrestrial broadcasting can more easily afford. With such a "technical" vision of broadcasting which does not consider the content of TV programs but only the customer's choice between alternative mediums, the broadcasting economical model should be reassessed to maintain fair competition between technologies and operators. As an example, the broadcasters should pay for their assigned frequencies, as the mobile operators do.

Let us come back now to the remarkable success of direct satellite TV which, after some years, has fully modified the broadcasting business. New important operators appeared which developed international strategies and gained some independence from national border lines and regulations, notably for the availability of their spectrum resources and the programs carried. However, this satellite TV broadcasting which is so common today, with receiving dishes spread everywhere, hesitated on a technical strategy, at the beginning, for spectrum allocation. The first planning conferences, notably a World Administrative Radio Conference, in 1977, made a clear distinction between bands for satellite broadcasting services and those for fixed satellite services, mainly fixed telecommunications services. The experts who represented these two parties stressed the differences between their services, not to conflict on common resources. Considering such a border line established between two separate worlds, the telecommunications operators, for example, overestimated the market for satellite business digital communications which was an important argument to obtain more bands. Simultaneously, the broadcasters made restrictive technical choices, notably improved analog modulations, which drastically limited their satellite capacity.

The market and the new international operators got rid of such directions during the 1990s and obviously demonstrated that services and techniques can converge on a single and simple objective: to provide all customers, within a broad geographic region, with hundreds of digital TV programs. Once again radio opened a new way to progress and showed that some dogmas were obsolete. Every year, now, throughout the world, more satellite TV transmitters (sometimes called transponders) are placed in geostationary orbit. Thanks to efficient picture codecs, it can be estimated that, today, 20,000 TV programs are broadcast by satellites to a billion customers, worldwide. In Europe, the companies EUTELSAT and SES-ASTRA are satellite broadcasting leaders. The geostationary orbital positions for their platforms are strategic and represent a major economic value. The longitudes 13°E for EUTELSAT and 19.2°E for SES-ASTRA are among the densest positions in the world for the use of bands 10.70 to 12.75 GHz.

It may be noted that, today, terrestrial TV and satellite direct TV are competing with a similar constraint: a directive receiving antenna oriented towards a nearby TV station or a geostationary satellite. This may not be the case in the future if terrestrial TV offers a portable or mobile service, at least in towns. It will not be so easily provided by satellites. Thus, it can be thought that complementary goals are attainable by these two broadcasting radio techniques: portability and mobility by one, capacity and universal coverage by the other.

Concerning satellite audio broadcasting, some dedicated companies, such as WORLDSTAR, have already been providing a service, around the world, since 1999. They offer direct reception of digital audio programs from geostationary satellites. Transmitting in the 1,467-1,492 MHz band, such satellites broadcast about 300 programs according to the MPEG3 standard. However, a constraint to getting the service is the use of an external and slightly directive antenna, which is not comfortable for mobile sets. This is a reason why it may be interesting to combine satellite and terrestrial infrastructures which provide a higher electromagnetic field. This cooperation between satellite and terrestrial audio broadcasting to DAB receivers looks like what was mentioned above on the possible complementary goals of satellite and terrestrial TV broadcasting, using the DVB standards. Field tests are needed to evaluate whether such scenarios are possible and attractive.

6.4. The promises of radiolocation

It is a recent idea that important markets may be attached to a permanent and precise knowledge of the location of people and things. For a long period, such services were considered professional and relevant to particular mobiles such as planes, ships, trucks or any heavy transportation means. Now they have become familiar to anyone for very different purposes and many bands, called aeronautical bands, are devoted to them.

Since the Second World War, radars have been able to evaluate the in-flight position of planes. From this basic radar principle, many different pieces of equipment have been designed for various needs, in all contexts, in the air, at sea, on the roads and are commonly used. Simultaneously, radio flight auxiliaries were also invented as VOR beacons (VHF omnidirectional range), in the aeronautical radiolocation band 108-117.975 MHz, to help direct planes in the airport vicinity, as well as the ILS (instrument landing system), 74.8-75.2 MHz and 328.6-335.4 MHz, or the DME (distance measuring equipment), 960-1,215 MHz. LORAN provides long distance location information.

Such detection and measurement systems used for radiolocation tightly combine radio transmission and signal processing techniques. The objective is not to transmit any message between transmitting and receiving stations, except identification data, but to analyze the received wave structure to find information concerning the position and movement of the concerned parties: transmitter, receiver, obstacles. Their performances progress steadily with the help of more accurate and sophisticated technologies.

Satellites, in a privileged position above the Earth, have created completely new opportunities. Apart from former terrestrial systems which were implemented for local purposes, worldwide satellite systems for radionavigation or radiolocation have appeared during the last 20 years, mainly for professional activities, but more and more for public use. Some examples follow.

The international network SARSAT was born in 1982 and was strengthened by an international agreement signed in 1988. It uses low orbit satellites which detect signals transmitted from beacons attached to mobile equipment at frequencies 121.5 and 406 MHz. The received signals are forwarded to earth stations which can estimate the position of the transmitting beacon with a precision of a few kilometers and identify it. Notably, in case of an accident and if necessary, teams can be sent into the field with the appropriate information to rescue any victim.

It is the GPS (global positioning satellite) system which made radiolocation services popular with simple and cheap terminals which indicate their position at any moment and anywhere on Earth, with a remarkably good precision. This system was created, at the beginning, for US Defense purposes but it has been made available to anybody with downgraded, but attractive performances, for all civilian applications, when the highest precision and information security are not strictly required. In 2000, the GPS network was established through 24 low orbit satellites in the 1,215-1,260 and 1,559-1,610 MHz bands, which were shared with the Russian system GLONASS. Since 1999, Europe has wanted to have its own system, GALILEO, and the US administration has wanted to improve the GPS performances. Thus, the World Radio Conference, Istanbul 2000, allocated new bands to satellite radiolocation services: 1,164-1,214, 1,260-1,300, 5,010-5,030 MHz and also 1,300-1,350, 5,000-5,010 MHz for satellite uplinks. Now, the resource can satisfy all the needs expressed. When these different projects are operational, they may replace most of the traditional instruments dedicated to radionavigation, with various service classes adapted to different needs, from the most secure to ordinary ones. It may be imagined that, in the future, most moving or portable equipment is equipped with a permanent radiolocation function, as common as time indication or personal identification, to contribute to countless applications such as car guidance, container supervision and so on.

Finally, let us mention again that some terrestrial systems, such as cellular radiotelephone networks, can also offer a radiolocation service with a limited precision.

6.5. Limits of the spectrum planning efficiency

What was mentioned above in this chapter about some major projects directed to mass market objectives, which are completed with many other systems dedicated to more specific uses, can raise questions about the efficiency of the existing spectrum planning practices. The best technical efforts, the most confident business prospects are only justified when the projects are achieved and found successful for customers and users. This is not always the case and setbacks are as common as successes. Considering the time scale of frequency planning decisions, which may be as long as several years (up to 10 years, sometimes), there are risks of making mistakes and wasting a scarce resource. In the same way, it can be found that such a long time to allocate bands to a new strategic program is unacceptable. Thus, some criticism has recently appeared against the traditional allocation and assignment process.

Good management should not constantly divide the spectrum to satisfy any project, notably with too local or specific purposes, which wants exclusive bands or fixed plans. On the contrary, the correct way is to look for a flexible sharing of common resources by different services and applications, having thoroughly studied the electromagnetic compatibility constraints and making the technical specifications of systems adapted to such constraints. With this policy, bands are available more quickly and the spectrum efficiency increases. Regulation and technology cooperate for a more fluent management. In the case where an application fails or does not meet the market, the corresponding bands are still used by compatible services.

The ITU has already set limits on the ever increasing spectrum "slicing" with the identification of a limited number of regulatory services which are not actual applications but general families where many different practical applications can take place. This opened a first stage for flexibility of the Regulations, avoiding considering the innumerable particular cases and the combining of them. Moreover, regulatory services are constrained by regulatory "hard limits" which are mandatory technical rules designed to increase the electromagnetic compatibility between systems. Finally, coordination rules and procedures are established, which should be used for a case-by-case approach, to assess the local compatibility of any actual radio station with others. These rules are sometimes called "soft limits" and still improve the flexibility.

Another merit of the ITU policy is to ignore the different national regulators dedicated to particular radio business, whose restricted views and responsibilities are brakes for a fluent use of frequencies. Every country gets only one representative Member which should debate for all questions and then has to promote arbitrations between its own spectrum users. As an example, the "convergence" rhetoric which is now expressed in the ITU or other international bodies, is not only commonplace for explaining that techniques and markets are unifying in the modern "information society". It also stresses that some artificial border lines still exist between different techniques and regulatory practices for obsolete reasons which may impair the global spectrum efficiency.

However, more opportunities appear, thanks to the "intelligence" and "agility" of radio networks and terminals. The dynamic spectrum access techniques, software defined systems, cooperative radio and the improved signal processors make possible new concepts for a more flexible assignment of frequencies and band allocation to services. The ideas of multi-purpose digital infrastructures (WAPECS) and frequency spread systems are also gaining support. They will be introduced and discussed in Chapter 10 to understand their capacities and limits.

It is not a demonstrated "theorem" but the best spectrum efficiency and flexibility are probably achieved when the spectrum power density is the most equally distributed over a widest band and a large area. Any density peak in the spectrum or on a geographical site provides a local advantage, technical or economical, to some particular system which creates this peak, but it degrades the global spectrum resource efficiency. Criteria for an improved management could be to "smooth" the radio power density parameters with more distributed applications. The new technical ideas but also the evolutions of traditional techniques and practices are aiming towards this objective.

Chapter 7

Regulatory Instruments for Spectrum Sharing

As sketched in Chapter 4, several regulatory instruments are available to insure a dynamic sharing of the radio spectrum between countries, services and users. The ultimate objective is the most efficient use of this common and scarce resource in a cooperative spirit which derives from a consensus upon a global strategy.

Experts should answer many demands every day, some of which are contradictory, and must simultaneously respect basic principles and requirements:

– to care for the existing operational services, protect them and let them grow, as necessary, or to foster them to decrease, if obsolete;

– to help innovative systems to appear and make possible the implementation of new applications;

– to let everybody enjoy the highest freedom of spectrum use within the limits of other user rights and freedoms.

Such activity and constraints should be managed in a fast moving domain. The spectrum is permanently reshaped to cope with the technological evolution, the economical and social life. The experts must be reactive and careful with various and appropriate tools.

The words "regulatory instruments" deserve some observations. They are rules and formal procedures which should be followed by all parties, being based on legal principles but also on deep technical analysis. They are mainly formalized at an international level, in the ITU Radio Regulations, but are also adapted at a national or regional level. They can really be considered a common reference for experts throughout the world.

When any new resource allocation is requested, a comprehensive study has to be made of the following items in the candidate frequency bands:

– a stringent evaluation of spectrum needs for the proposed service or application;

– an assessment of the electromagnetic compatibility conditions to be maintained in the service area with the existing networks and stations, and the necessary coordination procedures;

– the technical specifications and constraints to be satisfied by any new system.

This evaluation is a prerequisite before any spectrum sharing adjustment, at any level. The study results should be compared with the basic requirements, as stated above:

– What impact on existing operational networks?

– Is the new offer attractive enough and does it deserve spectrum management efforts to be implemented?

– Are the technical specifications conveniently adjusted for the best balance between constraints and freedom?

Care should be taken of the actual and various situations, in the field, and of the different legitimacy of partners and services, no one being thought negligible:

– international legitimacy of states with a request for democratic equity between them, having regard to different needs and economical situations;

– legitimacy of a concurrence between services and operators, taking into account their different social values: business, public security, science, defense, etc.;

– legitimacy of existing actors, their investments and their current operational activity, when compared to prospects and long-term benefits.

Such detailed analyzes generally result in rather conservative opinions. Spectrum managers do not make important changes unless a careful study has shown that they will not seriously impair any partner. Common decisions are taken from their consensus. Thus, the international regulations and their national derivatives have been built, since the beginning of the 20th century, on balanced approaches, each step being a compromise between actors.

However, a fundamental principle can be found from past experience, which is not clearly stated in the regulations but is practically ensured, which says: "first come, first served". In the absence of precise contradictory rules, if in conformance with current regulations and at any time, a user who wants to implement a service, to install a radio system or a new station is allowed to do so. Furthermore, having registered his project, he gains the benefit of rights which can be opposed to any other party coming later. Depending on the regulatory status of the service, the new application radio links will be protected, in the future, against any interference which may happen.

The most common application of this principle concerns frequency assignments to operators for stations. This will be developed in the next chapter and procedures will be described which guarantee the corresponding rights. The assignments are filed in regulatory data bases, either at the ITU or at a national level.

The same philosophy applies to promoting new services and maintaining the rights of existing ones when planning the spectrum. At an international level, the spectrum organization results from successive arrangements decided by periodical world conferences. Generally, each arrangement tries to minimize its own impact on existing situations. Thus, the improvement of the global spectrum efficiency proceeds from detailed adjustments of the electromagnetic compatibility between services rather than global movements or new stringent technical constraints. In fact, such an "authoritative" policy could not be adopted by any international organization because it would harm too many existing rights and thus could not obtain a consensus.

The main regulatory tools to share the spectrum resource are the following:

– the frequency band allocation tables;

– the plans;

– the coordination procedures before individual frequency assignments.

These instruments are complemented with technical rules and specifications which limit the characteristics of radio systems to guarantee the spectrum sharing efficiency. These rules are "hard limits" for the development of networks and stations to confine waves at a maximum. They add to the "soft limits" based on coordination of assignments which are negotiated after a case-by-case study between the parties concerned.

7.1. Frequency allocation tables

It was explained in Chapter 1 that the main and simplest way to share the spectrum resource is to allocate different frequency bands to services and users because the selectivity of the receivers is a very easy and straightforward tool to separate waves.

The frequency allocation tables describe how the whole spectrum is shared between services, being divided into frequency bands. They are issued at two main different levels:

– the international level of the ITU with the frequency allocation table to the different regulatory services, which is an essential part of the Radio Regulations;

– the national level. Every country should have a national table which is not only a local adaptation of the ITU table to a particular context, with specific choices among the allowed regulatory services, but which gives all the appropriate information for practical use of the spectrum resource in its territory by operators, for designed applications.

To introduce such tables, the best way is to present and comment on a short extract.

2,170-2,520 MHz

Allocation to services		
Region 1	**Region 2**	**Region 3**
2,450-2,483.5 FIXED MOBILE RADIOLOCATION 5.150 5.397	**2,450-2,483.5** FIXED MOBILE RADIOLOCATION 5.150 5.394	
2,483.5-2,500 FIXED MOBILE MOBILE-SATELLITE (space-to-Earth) 5.351A RADIOLOCATION	**2,483.5-2,500** FIXED MOBILE MOBILE-SATELLITE (space-to-Earth) 5.351A RADIOLOCATION RADIODETERMINATION-SATELLITE (space-to-Earth) 5.398	**2,483.5-2,500** FIXED MOBILE MOBILE-SATELLITE (space-to-Earth) 5.351A RADIOLOCATION RADIODETERMINATION-SATELLITE (space-to-Earth) 5.398
5.150 5.371 5.397 5.398 5.399 5.400 5.402	5.150 5.402	5.150 5.400 5.402

Corresponding footnotes

5.150 The following bands:

13,553-13,567 kHz	(center frequency 13,560 kHz),
26,957-27,283 kHz	(center frequency 27,120 kHz),
40.66-40.70 MHz	(center frequency 40.68 MHz),
902-928 MHz	in Region 2 (center frequency 915 MHz),
2,400-2,500 MHz	(center frequency 2,450 MHz),
5,725-5,875 MHz	(center frequency 5,800 MHz),
24-24.25 GHz	(center frequency 24.125 GHz)

are also designated for industrial, scientific and medical (ISM) applications. Radiocommunication services operating within these bands must accept harmful interference which may be caused by these applications. ISM equipment operating in these bands is subject to the provisions of No. **15.13**.

5.351A For the use of the bands 1,525-1,544 MHz, 1,545-1,559 MHz, 1,610-1,626.5 MHz, 1,626.5-1,645.5 MHz, 1,646.5-1,660.5 MHz, 1,980-2,010 MHz, 2,170-2,200 MHz, 2,483.5-2,500 MHz, 2,500-2,520 MHz and 2,670-2,690 MHz by the mobile-satellite service, see Resolutions **212 (Rev.WRC-97)** and **225 (WRC-2000)**.

5.371 *Additional allocation:* in Region 1, the bands 1,610-1,626.5 MHz (Earth-to-space) and 2,483.5-2,500 MHz (space-to-Earth) are also allocated to the radiodetermination-satellite service on a secondary basis, subject to agreement obtained under No. **9.21**.

5.394 In the USA, the use of the band 2,300-2,390 MHz by the aeronautical mobile service for telemetry has priority over other uses by the mobile services. In Canada, the use of the band 2,300-2,483.5 MHz by the aeronautical mobile service for telemetry has priority over other uses by the mobile services.

5.397 *Different category of service:* in France, the band 2,450-2,500 MHz is allocated on a primary basis to the radiolocation service (see No. **5.33**). Such use is subject to agreement with administrations having services operating or planned to operate in accordance with the Table of Frequency Allocations which may be affected.

5.398 In respect of the radiodetermination-satellite service in the band 2,483.5-2,500 MHz, the provisions of No. **4.10** do not apply.

5.399 In Region 1, in countries other than those listed in No. **5.400**, harmful interference shall not be caused to, or protection shall not be claimed from, stations of the radiolocation service by stations of the radiodetermination satellite service.

5.400 *Different category of service:* in Angola, Australia, Bangladesh, Burundi, China, Eritrea, Ethiopia, India, Iran (Islamic Republic of), the Libyan Arab Jamahiriya, Lebanon, Liberia, Madagascar, Mali, Pakistan, Papua New Guinea, the Dem. Rep. of the Congo, the Syrian Arab Republic, Sudan, Swaziland, Togo and Zambia, the allocation of the band 2,483.5-2,500 MHz to the radiodetermination-satellite service (space-to-Earth) is on a primary basis (see No. **5.33**), subject to agreement obtained under No. **9.21** from countries not listed in this provision. (WRC-03).

5.402 The use of the band 2,483.5-2,500 MHz by the mobile-satellite and the radiodetermination-satellite services is subject to the coordination under No. **9.11A**. Administrations are urged to take all practicable steps to prevent harmful interference to the radio astronomy service from emissions in the 2,483.5-2,500 MHz band, especially those caused by second-harmonic radiation that would fall into the 4,990-5,000 MHz band allocated to the radio astronomy service worldwide.

Figure 7.1a. *Radio Regulations (ITU, version 2004).*
Extract from the frequency allocation table. Band 2,450-2,500 MHz

Let us consider the band 2,450-2,500 MHz. At the border of the UHF and SHF bands, it is nowadays a convergence domain which can be used by both the fixed service [FIXED] and the mobile service [MOBILE]. Both services are authorized by the Regulations, in all Regions, with a primary status.

This band has been used for a long time by infrastructure long distance links (terrestrial fixed service) which are progressively replaced by mobile applications. For example, radio local area networks, RLANs, such as WiFi (mobile service), are now authorized. It is thought that before 2010, the UHF band will stop being used for fixed services, at least in the most developed countries. Typically, the 2,500-2,690 MHz band is already reserved for the UMTS future extension in Europe. In some countries, it has also been preferred to implement local TV broadcasting services in the 2,450-2,500 MHz band with so-called MMDS equipment. Normally, this band is not allocated to broadcasting but MMDS technology looks very much like a fixed service and can be accepted for such a local need.

In the same band, the first satellite networks for mobile telecommunications established their down links towards mobile terminals, with the provision of the MOBILE SATELLITE (space-to-Earth) service allocation. Furthermore, military ground-to-air radars are installed in some countries according to the RADIOLOCATION allocation.

All these services and applications can share the same spectrum if particular technical rules are specified and enforced, depending on country choices. This is the reason why they appear in the frequency allocation table of the Radio Regulations, as shown in Figure 7.1a. Let us comment more deeply on this extract.

Two bands are distinct: 2,450-2,483.5 MHz and 2,483.5-2,500 MHz. The difference is that space service downlinks, MOBILE SATELLITE and RADIODETERMINATION SATELLITE, are allocated in the second band and not in the first. This is a common situation. When two low directivity applications, such as terrestrial mobile service and satellite mobile service, are allowed in the same band, it is generally necessary to divide this band into two sub-bands which are allocated to the different services, to avoid reciprocal interferences. A footnote 5.351A gives some indications about the appropriate conditions to use the second band by third generation mobile satellite networks (IMT2000).

Some differences can be noted and explained between Region 1 (Europe, Africa, Middle-East) and the other Regions. In Regions 2 and 3, radiolocation is allowed with a primary status but it benefits only from a secondary status in Region 1. This difference is marked by the typography: RADIOLOCATION, with capital letters, in

Regions 2 and 3, radiolocation, with small letters, in Region 1. Let us recall how a secondary service is considered compared to a primary service:

– a secondary service should not interfere with a primary service, even when the primary service comes later;

– a secondary service cannot request any interference protection from a primary service, even coming later;

– priority rights may exist between secondary services.

A footnote, 5.397, indicates that France is allowed to give primary status to radiolocation, on its territory, as an exception in Region 1.

Another difference between Regions concerns radiodetermination satellite services. They are not allowed in Region 1, primary in Region 2, secondary in Region 3. However, national agreements are negotiated on this matter (see footnote 5.400).

In both bands, another footnote [5.150] authorizes the industrial, scientific and medical applications (ISM). Such applications are not radiocommunications but local use of electromagnetic waves for various purposes, such as heating. The ISM waves are normally confined and should not be propagated outside the confinement area, but radio users, in the vicinity, should take care of possible local disturbances created by such equipment.

Finally, it should be remembered that satellite services in these bands must apply coordination procedures [5.402].

This short example shows the precision of the Regulations but also that many choices stay open, within well defined limits. However, it concerns only a tiny part of the spectrum, 50 MHz wide. The whole usable spectrum is organized in the same way, according to thousands of rules, spread over several hundred bands and sub-bands. Practically, the allocation table included in the Regulations is divided into almost a hundred pages with an equivalent volume of footnotes for more information, notably for accepted national exceptions.

On this regulatory international basis, each country can build its own allocation and utilization table. These tables are very different, more or less detailed. They depend on national legal provisions and take into account regional harmonization directions. However, their main features are very similar to the ITU tables:

– bands and sub-bands must respect the ITU structure, with some complementary divisions, if necessary, to share the resource according to local needs;

– particular regulators and authorities may be designed to manage the corresponding resources. a rather common distinction is made between defense, governmental and commercial bands;

– choices are made between the possible services authorized by the ITU regulations and more precise applications designated;

– priority status can be more elaborated;

– any useful administrative, technical or geographical specification or indication may be added.

The ERO (European Radiocommunication Office), a permanent technical office of CEPT, publishes a European Common Allocation (ECA) table which results from the harmonization debates between CEPT members and which may be considered as a common objective for a future European integrated spectrum. It is, of course, not so detailed as national provisions but it summarizes the consensus achieved by the different countries. In addition to this ECA table, ERO publishes a summary of every national table from the different European countries. All together, this EFIS (ERO frequency information system) site gives valuable information on the spectrum use in Europe.

Frequency Range: 2450   to 2500   MHz ▾   Frequency Table: - Europe (ECA) - ▾   Search

Results from the ERO Frequency Database:

FREQUENCY BAND	ALLOCATIONS	APPLICATIONS
2450.0 - 2483.5 MHz	FIXED MOBILE	Detection of movement (2400.0 - 2483.5 MHz) Non-specific SRDs (2400.0 - 2483.5 MHz) RFID (2400.0 - 2483.5 MHz) Radio LANs (2400.0 - 2483.5 MHz) AVI (2446.0 - 2454.0 MHz) ISM
2483.5 - 2500.0 MHz	FIXED MOBILE MOBILE-SATELLITE (space-to-Earth)	Fixed links ISM Land mobile MSS Earth stations SAP/SAB and ENG/OB

Figure 7.1b. *CEPT harmonized table. Extract from the ECA table as published by ERO*

Let us note that the ECA table is restricted to Region 1 and considers only primary allocations. For the 2,450-2,500 MHz band, it is fully compatible with the ITU allocations. ECA has made the choice to dedicate the 2,450-2,483.5 MHz sub-band to different short range devices (SRD), notably RLANs, which are considered mobile services. Choices remain more open for the 2,483.5-2,500 MHz sub-band.

It seems obvious that with so many detailed provisions, notably at the ITU level, the rights of users may be sometimes difficult to assess. Experts are needed to interpret the rules and arbitrate conflicts, if any. In every country, an administrative expert team should be appointed to this task. It should notably analyze the user requests for frequency assignments and give advice about them. When an assignment is given, priority rights are attached. They can be registered at a national and ITU level. Thus, in the ITU, an international expert committee is convened on such arbitration matters, when necessary, the RRB (Radio Regulation Bureau). When no difficulty appears, the assignment can be dealt with through administrative procedures by the Radiocommunication Bureau which registers it in the Master International Frequency Register (MIFR).

7.2. Plans

The frequency allocation tables are general all-purpose instruments which leave open many choices and opportunities for national actors to implement real applications, install new stations and innovate, when necessary. Another important tool does exist which is adapted to major and sensitive services if stable enough, internationally harmonized and for long-term use: plans. Of course the allocation tables and plans are fully compatible.

A plan describes the reference organization (geographical and spectral organization) of a particular radio application, on a worldwide, regional or national area, perhaps on a smaller territory, according to *a priori* agreements between the parties concerned.

Most generally, the spectrum resource, a frequency band, is divided into standard channels (see Chapter 2, section 2.1). The geographical area may also be divided into delimited parts. A plan should take care of existing stations with their associated rights and future incoming ones: all of which must find appropriate channel resources to operate in the plan, without uncertainties. This instrument is thus well adapted to a global and smooth development of a common harmonized service throughout countries, even with different situations, notably economical levels, with a fair equity between them. Plans contribute to creating, over large areas, standardized and stable public services, such as broadcasting services, or worldwide security services such as aeronautical or maritime networks. They are seen as an appropriate answer to politically sensitive national concerns since equitable rights of the different administrations are protected, from the beginning. A plan is formally registered at an appropriate level and any modification must be negotiated. Unless such complementary discussions occur, future channel assignments must fully conform. Some plans are part of the Radio Regulations, while others are managed at a regional level.

A plan may be an assignment plan where any station is considered, on a case-by-case basis, and its electromagnetic compatibility assessed with all existing ones. The decision to build such a plan is guided by an already operational situation when the plan is designed, with many working stations having previous registered rights. At that time, adaptations may be negotiated to ease future developments. Any change or evolution which comes later on a planned station must be studied, notified and registered.

A plan may also be an allotment plan where the spectrum resource is divided and shared, from the beginning, to "tile" uniformly the territory concerned and avoid any interference, in the future. This scheme is better adapted to a new network deployment. In such a prospective plan, the spectrum resource is allotted. This means that it is shared from the beginning between the participating countries, sometimes with precise geographical provisions, in such a way that each one gets a frequency lot adapted to its present and future needs, on an equitable basis. The plan may not only provide each country with a global frequency allotment, it may describe a precise distribution of channels or sub-bands over definite areas (as big cells), which makes possible the best frequency reuse among the participating countries. However, some more resources or exceptions may be requested by any party apart from the plan provisions. If the plan is managed by the ITU, a demand should be transmitted to the Radiocommunication Bureau which studies it and makes any necessary enquiry with the concerned administrations to get advice and agreement. A formal procedure is applied to back this plan adaptation. If accepted, it will be registered as part of the plan and give new rights and limits to the parties.

A typical example of a plan, probably the oldest, since radiocommunications with ships began to be internationally organized at the beginning of the 20th century, notably after the Titanic disaster (1912), is described in Appendix S25 of the Radio Regulations. It is entitled: Provisions and associated frequency allotment plan for coastal radiotelephone stations operating in the exclusive maritime mobile bands between 4,000 kHz and 27,500 kHz. This service, in the HF band, although rather obsolete in Europe, is still much used in many parts of the world. Different systems are dedicated to maritime communications and several to maritime mobile communications, using various sub-bands:

– 4,063-4,438 kHz, so-called 4 MHz band;

– 6,200-6,525 kHz, so-called 6 MHz band, etc.

Among these bands, some are used for duplex telephony and are the subject of the S25 appendix. Waves are amplitude modulated (single sideband modulation, SSB) and occupy 3 kHz wide channels. Some channels are dedicated to ship transmission, others to coastal station transmission. Each one is allocated to several particular countries.

Several other plans are described as an appendix to the Radio Regulations. The APS26 and APS27 plans concern the aeronautical mobile service in HF bands. Notably, APS27 is dedicated to in-flight civil aeronautic communications between planes and the ground. The corresponding channels are also 3 kHz wide.

Such plans seem to be the most necessary. They concern transportation safety and must be internationally managed since planes and ships can travel anywhere on Earth and should be supplied with standardized radio equipment. However, for years, other services appeared which were thought economically or politically sensitive and thus should also be planned. This is notably the case for satellite systems. The purpose is to maintain equity between countries to access a precious and scarce resource: the geostationary satellite positions for the different frequency bands allocated to satellite services. These positions are used to place satellites which stay fixed above the Earth and thus are ideal radio relay platforms to permanently cover a wide geographical area. Every state government estimates that it is strategic to control several orbital places above its territory and thus obtain particular rights on the corresponding positions and frequency bands, in order that the whole resource is not monopolized by the most advanced countries. This is the reason for an allotment plan.

Among the most debated plans are those described in Appendix 30 of the Regulations, which concern the broadcasting satellite service for direct TV to customers through individual parabolic antennae. They organize the bands 11.7-12.5 GHz in Region 1, 12.2-12.7 GHz in Region 2 and 11.7-12.2 GHz in Region 3. The planning studies were initiated at a World Administrative Radiocommunication Conference (WARC) in 1997 and many subsequent meetings improved the document. Some important contributions were discussed at the World Radiocommunication Conference (WRC), Istanbul 2000, where resources were allotted to every country in Regions 1 and 3: 10 TV channels per country in Region 1, 12 in Region 3.

With the same view of equity between countries, a twin plan has been defined for uplinks of the broadcasting satellite service. The corresponding bands are 14.5-14.8 GHz (outside Europe), 17-3-18.1 GHz in Regions 1 and 3, 17.3-17.8 in Region 2.

Let us also mention Appendix 30B which partly plans the fixed satellite service. It concerns the bands 4.5-4.8 GHz, 10.70-10.95 GHz, 11.20-11.45 GHz for the downlinks from space to the Earth, and 6.725-7.05 GHz, 12.75-13.35 GHz for uplinks. It includes national allotments but also existing networks with former rights. In this plan, a lot is characterized by a set of different items:

– a nominal geostationary orbital position (longitude);

– a national service area, described as an ellipse at the Earth's surface;

– some technical parameters concerning the antenna radiation pattern;

– a predetermined longitude arc on the geostationary orbit which can be used as a margin to eventually move the satellite from its nominal position, if the orbital places should be modified.

Normally the whole band can be accessed by any administration, ITU member. The angular distance of the orbital positions and the geographical distances between stations provide a sufficient guarantee against interferences and make it possible for any system to use the full spectrum resource. Thus, this appendix actually describes really an orbital position allotment plan.

Some other plans are regional. This is typically the case for terrestrial broadcasting. One of the best known plans is Stockholm 1961, adapted by Geneva 1989, which replaced a former Stockholm 1952 to organize the main broadcasting bands in Europe and Region 1. The bands are the following:

– Band 1   47-68 MHz

– Band 2   87.5-108 MHz

– Band 3   174-230 MHz

– Band 4   470-582 MHz

– Band 5   582-960 MHz

Band 2 concerns FM radio. The upper bands, 3, 4 and 5, concern TV, notably the UHF bands, 4 and 5, which are divided into 61 8 MHz wide channels, numbered from 21 to 81. Every channel carries an audio and video signal. The plan has been recently modified by two Regional conferences, Geneva 2003 and Geneva 2006, to convert analog TV to digital modulations (see Chapters 5 and 6), keeping the basic 8 MHz channel organization, each one now carrying a digital multiplex. It provides an international framework to deploy digital TV and radio services with the appropriate frequency allotment for countries in Europe, Africa and the Middle East. To ease the transition from analog to digital modulation in border regions, a coordination procedure has been enforced.

Another recent broadcasting regional plan is called Wiesbaden 1995-Maastricht 2002 which allots between European countries the frequency bands allocated to the new digital audio broadcasting system, TDAB, notably the 1,452-1,492 MHz band which is shared between terrestrial and satellite broadcasting services.

Maastricht 2002 Plan

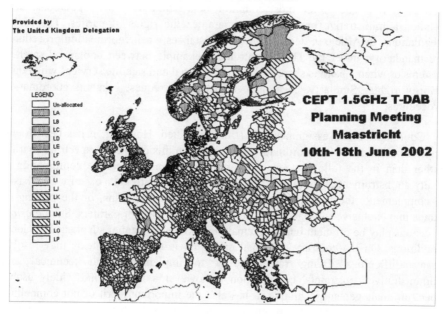

Figure 7.2. *Allotment for the TDAB application in Europe, band 1,452-1,492 MHz. Plans organize the spectrum resource over a wide area, for a particular service. At Wiesbaden, in 1995, thus at Maastricht, in 2002, European countries decided to plan the new broadcasting service TDAB. The plan allots the spectrum resource into channel lots which are geographically distributed*

There are also local plans and arrangements which are discussed and issued by neighboring countries in order to harmonize their network deployments in border zones. One of the main features of such local arrangements is *a priori* band sharing in these areas with the designation of preferential sub-bands allotted to every participating country. In its preferential sub-band, a country can install stations without any previous negotiation with neighbors.

7.3. Coordination

To decide the exact frequency or channel for a new station or radio link in conformance with the regulations, to authorize this station to transmit without producing interferences to any other existing one, to grant it priority rights and to register them in regulatory files means to assign a spectrum resource. This procedure will be studied in Chapter 8.

An important step during the assignment procedure is the coordination, when necessary. It is one of the most common tools required by the regulations, with detailed administrative provisions, to guarantee the rights of parties. The Radio Regulations, in Article 9, describe the circumstances when coordination procedures are mandatory, under the ITU's control, for example between unplanned satellite systems or when a band is shared between terrestrial and satellite services with equal priority rights. Similarly, local agreements often request bilateral coordination procedures.

Coordination and assignment are strongly related. However, it has been found appropriate to deal with coordination procedures in this chapter on spectrum sharing rather than in the following one concerning assignments because coordination is really an instrument to share the spectrum resource fluently on a case-by-case basis. It complements the frequency allocation to services, the plans, or the sharing of bands into exclusive sub-bands allocated to different users or partners. Such *a priori* decisions may be efficient but they remain inflexible and cannot adapt easily to local conditions. On the contrary, a formal dialog on in-the-field contexts may resolve many difficulties. Being rather heavy to implement, with technical and administrative constraints, coordination is however a flexible and widely usable spectrum management technique. It has also the important merit of not compelling regulators to "slice" the spectrum indefinitely into dedicated bands and channels which create stiffness and inefficiency, but to leave open opportunities for local cooperation between various users and services to share a common resource. The basis of the dialog is a case-by-case study of the relevant needs and constraints of participants to assess their mutual compatibility.

Any new station which is installed enters an existing radio "landscape" or environment. The new entrant should take care that, when transmitting, it will not interfere with existing links which share the same frequency band and have the advantage of former rights. This is made by calculations of permissible or accepted interferences. If the interference level remains lower than some reference values, the new station can be allowed to transmit. If the level exceeds these values, the new station cannot be allowed. It is the responsibility of the new station operator to assess the probability of interference with the stations which are already at work, with registered rights. From such investigations, he can decide on the practicability of his project.

International coordination procedures often mention the possibility of designing coordination areas around existing receiving stations, notably earth stations. They are calculated from their technical characteristics, taking into account the interference evaluation from transmitting stations which would be installed later.

A coordination area is such that any new transmitting station which may be installed outside will not create harmful interferences to the existing station concerned. Thus, if an operator builds a new station outside this area, he is allowed to put it into service without any previous coordination procedure because no interference conflict is forecast. On the contrary, if a new station is going to be installed inside the coordination area of an existing station, or when such an area does not exist, a coordination procedure is necessary.

Coordination is very thoroughly organized and formal. The administration which backs the new project must produce a standardized technical document where the station parameters are described and transmit it to the other administrations concerned with existing stations in the same frequency band. These administrations have a limited time to study, comment, agree, refuse or raise objections to the project. The transmit authorization directly depends on these answers which should be based on demonstrative technical arguments. If any interpretation difference occurs, ITU expertise may be asked for. In the case when no agreement is obtained, a probative operation of the new station may be decided with controversial parameters. If, during a few months, no interference is observed, the new station can be formally approved and registered. On the contrary, if disturbances occur, it should be stopped. All steps of this enquiry are carefully described in the Regulations.

Let us briefly mention some typical situations, as examples.

Article 9 of the Regulations describes the different configurations where coordination is necessary. For example, section 9-18 analyzes the common case of a fixed terrestrial service station, such as a microwave link station, which transmits in the same frequency band as an earth station, part of a satellite telecommunications network. It establishes the conditions which make coordination mandatory. Appendix 5 indicates which administrations should be notified. Appendix 7 describes a method to draw the coordination area around an earth station and gives the basis for the calculation of possible interferences. The ITU keeps a file where the main earth station coordination areas are described, based on these methods.

The international coordination of short distance links, such as fixed service microwave links or terrestrial mobile stations, is mainly decided at a local level according to bilateral or regional agreements which are negotiated between neighboring countries or within regional forums, notably to make the installation of stations in the border areas easier. In Europe, many countries have signed such coordination agreements, the most well known being Vienna 1986 which has been modified many times and is now called the Berlin agreement. A new version was approved in 2000 by 16 countries to coordinate radio links between 29.7 MHz and 43.5 GHz for the fixed service and base stations for the terrestrial mobile service.

This agreement describes the bands to be considered, the necessary technical parameters to calculate the radio field values which do not create harmful interferences, the harmonized calculation method (HCM) to evaluate the actual fields and the coordination administrative procedures. A reference digital map of Europe has been chosen and provided to all partners to obtain common calculation results. An interesting instrument has been defined which is an admissible interfering field level at the border lines, which can replace the coordination area. Any station installed in a country which does not create at its border a radio field with intensity higher than this reference value can be activated without international coordination.

Many other arrangements have been approved between European countries to harmonize and fasten the deployment of different networks in some bands, notably for mobile networks with the GSM or UMTS standards. They include the definition of preferential sub-bands in border zones to avoid case-by-case station coordination and give more freedom to the operators.

The satellite systems are particularly concerned by international coordination procedures because orbits and orbital positions are strategic and scarce resources which, by their own nature, appear to be international goods. They interest all nations and must be controlled by the ITU. Apart from some services which are planned (see section 7.2), many others remain open to competition. The corresponding space and radio resources (orbits and frequencies) can be obtained after regulatory coordination and assignment procedures based on "paper projects" and become strategic advantages for the future. Here the basic principle, "first come, first served", fully applies and may produce unfair results due to speculative space resource provisions. The Radiocommunication Bureau of the ITU plays a major role, as a central point to deal with information concerning unplanned satellite systems and to insure that all appropriate procedures are conveniently applied. Notably, this bureau publishes all projects where a satellite appears which may interfere on other networks, or be interfered with.

Article 9 of the Regulations, already mentioned, describes all coordination circumstances for satellite systems. There are many, with various combinations: geo and non-geosatellites, terrestrial and earth stations, planned and unplanned assignments, etc. Let us consider, as an example, the case described in section 9-7, concerning the coordination between assignments of two unplanned geostationary satellite networks. A new satellite must demonstrate that it will not harm existing services. One of the main items of proof to be provided is the noise temperature increase, ΔT, that the new satellite will create in the receiving stations of other networks (see Chapter 2, sections 2.2 and 2.3). The relative noise temperature increase $\Delta T/T$ should remain lower than 6%. Appendix 8 describes the corresponding calculation methods.

All the above coordination techniques and procedures are mainly set for international contexts. However, they can also be used, in a less formal way, for national needs, for example between operators or administrations which share a common band. It is the responsibility of national regulators to establish the corresponding technical and administrative provisions.

7.4. Technical limits

Networks are not completely free to use their assigned frequencies without constraints, even if they do not disturb any other station when they are put into service. The reason is that the spectrum, as a scarce resource, must be spared as much as possible to get its highest efficiency for all operators and services. Thus, many technical rules apply to confine the waves and limit future interferences when the same frequencies will be used by other systems. These rules concern the transmitted power (or EIRP), the transmission time cycle, the antenna radiation pattern, the pointing direction, the location of stations, the channel and modulated carrier spectrum characteristics, the out-of-band spurious signals, the tolerances on various radio parameters, etc.: any feature can be considered which may influence the electromagnetic compatibility with other systems. Together with state-of-the-art equipment characteristics, they protect radio links one from another. They back and guide the technical specifications of stations and networks.

It is not possible to discuss so many rules, some of which are very simple, some others complex. Let us only give a few examples for satellite services.

Article 21 of the Regulations gives some engineering rules which apply to earth stations and terrestrial stations to let them use the same frequency bands. With the idea that such a common and simultaneous use is possible by an angular distance between two radio beams at different elevations, it requests that such a distance be maintained between horizontal beams and beams oriented towards the satellites. The different parts of this article consider the various radio beams which can be found in such networks. Some rules follow:

– the EIRP (see Chapter 1, section 1.4) of the terrestrial fixed service stations should not exceed 55 dBW and the corresponding wave beams must keep distant from the geostationary orbit with an EIRP limited to 47 dBW when nearer than 0.5° from this orbit;

– the earth stations should not transmit with an elevation lower than 3°;

– the satellites should not radiate a too intense energy towards the Earth, notably with a horizontal incidence. The power flux density at the Earth surface should not exceed –52 dBW/m² per 4 kHz band for an incidence angle lower than 5° in the 3.4-4.2 GHz band.

Such technical rules which apply *a priori* to radiocommunications are sometimes called "hard limits". They will multiply and get more precise to increase the spectrum efficiency for new systems. They must also question the existing equipment and services for up-dating their characteristics. Although they enforce important and sometimes costly changes, they are necessary when they open strategic perspectives. As an example, this was the case when such rules made possible the sharing of common bands between terrestrial fixed services, geo and non-geosatellite services, as a conclusion of debates opened since the beginning of the space era.

It can be thought that a liberalization to access the spectrum resource which should limit the need of detailed band allocations and individual frequency assignments with previous coordination ("soft limits") will necessarily increase the need of "hard limits" which confine waves more strictly, from the beginning.

Chapter 8

Frequency Assignment: A Contract

The different regulations, notably the Radio Regulations, issued by the ITU, and the various frequency allocation tables, either national or international, build a large framework for spectrum utilization by services and are guides for regulators. They put limits on every circumstance to maintain the best efficiency of the radio resource, from a balance between operator rights and duties and contradictory technical parameters of systems. Complementary instruments, such as plans, organize some services on particular geographical areas more strictly. More tools are available, such as coordination procedures, to share fluently common frequency bands among neighboring users.

This regulatory framework can be considered spectrum common law. However, generally, spectrum usage is not freely open to anyone. It is a privilege of state administrations to deliver individual authorizations, licenses for frequency utilization, which are real contracts. A frequency license or administrative authorization allows a particular operator to use a spectrum resource, commonly a band or channel, for a designated radiocommunication service, over a precise geographical area and for a limited time, with some various conditions.

This act means a formal contract between the operator (or user) and the administration, being based on technical, legal and economic provisions which set out the rights and duties of both parties. Notably it can attribute a spectrum resource and assign precise frequencies to radio stations.

In a restrictive meaning, assignment is a standard procedure which is described by the Radio Regulations to allow a designated radio station to use a particular frequency. This assignment may be notified to the ITU Radiocommunication Bureau to be registered in a regulatory data base, the Master International Frequency Register (MIFR), which provides internationally recognized priority rights.

However, there is no objection, in common language, to enlarging the assignment meaning to the technical contract between spectrum users and managers and to make it include the full relation based on spectrum regulations, as a part of a global license or administrative authorization. In fact, spectrum utilization is not only ruled by international or national provisions directed towards the best spectrum efficiency; an operator also has to comply with more general legal constraints which impact particularly radio services and equipment. They may be relevant to public health, environmental concerns, social objectives, etc. Some more important constraints are designated as "essential requirements". Wave utilization also implies a full respect of these external constraints.

8.1. Contracting parties

As explained in Chapter 4, the most common view of state administrations is that the radio spectrum, as public goods, should be managed by the government (or any designated independent body) for the benefit of all users in the national territory, in compliance with the ITU recommendations and regulations. Every administration publishes adapted national laws and regulations to back up this management, notably a national allocation table which gives a detailed description of the possible spectrum usages in the country.

Among others, two user categories may be distinguished, those which use the spectrum for governmental purposes and those which use it for private, business, commercial purposes, or so considered. Depending on these categories, a national frequency assignment may look different, but most rules apply in both cases.

In the first situation, a national administration, being also the spectrum manager, retains part of the resource for its own needs. Thus, it is not necessary to set up a formal contract between administrative parties, the same government being simultaneously manager and user (even when these parties are different ministries inside the same national administration). However, it seems very useful that all common administrative and technical procedures be followed, notably frequency assignments to radio stations with the appropriate filings in national or international data bases. This is convenient for good technical management of the spectrum, as a whole, no part of it and no user being fully independent from the others. This is absolutely necessary if the government rights, as a spectrum user, should be considered and respected among foreign countries, notably in border regions or for satellite systems. This point will be further discussed later but it is clear that a governmental network cannot request any privilege from the international rules, outside the national territory. The administrations, as users, must comply with the common international procedures and practices to back up their own rights.

In the second situation, the administration provides a different party, like a private person, a business company, a public institution, with a share of public goods. This attribution of a public resource is commonly the matter of a license or authorization where rights and duties are defined, as a contract. Previous administrative and technical procedures may be necessary before issuing this authorization. It may be individual, as a personal act. It may also be implicit within a general public provision or a class license.

It may be recalled that any spectrum attribution or frequency assignment should be in line with the regulations, notably international, the regional agreements with neighboring countries, the frequency allocation tables and plans, etc. When assigning a frequency channel or attributing a spectrum resource, an administration is responsible for any interference which could appear due to contradictions with such documents.

8.2. Common bands and assignments

Some bands are available for particular applications without individual assignment. They are considered collective goods which radio stations can use freely, however respectful of technical and administrative rules. The corresponding specifications should be published and terminals, on the market, should comply. All together they shape an implicit contract which any user must accept.

More and more systems, mainly mass market products, are designed to be used in such conditions and more bands allocated for them. Some examples follow:

– citizen band (CB), familiar to car drivers for short distance communications on the road (27 MHz in Europe);

– SRD (or LPD) bands, used by mass market low power equipment such as telealarms, telecommands, cordless phones, etc. Convenient narrow bands are spread all along the spectrum for these terminals;

– local area network bands (RLAN) for equipment as WiFi, Hiperlan, Bluetooth, etc.

As already mentioned (see Chapter 5, section 5.1), the corresponding bands are generally allocated to mobile services, considering that the location of the equipment is not known. Such common bands may be largely used in the future to increase the flexibility of spectrum: Chapter 14, section 14.2 describes some prospects of this. It may be noted that the usage of common bands is normally free of charge since, generally, the users are anonymous and cannot be charged for their frequency utilization.

A frequent feature of this equipment is that its international standard allows it to operate with loose parameters (power and frequency band notably) which must be constricted when used in a particular country. This situation may create conflicts between product marketing and utilization if not properly managed. In some extreme cases, a product can be freely sold and bought in a country but should not be used on its territory.

Some other bands are usable for specific services without individual assignments, but the users must be individually licensed, sometimes after an examination, and must comply with particular operational rules which are parts of the Radio Regulations. Two examples are given:

– amateur bands, throughout the spectrum;

– maritime and aeronautical bands, usable notably for emergency calls or navigation help.

These bands are commonly harmonized worldwide.

A counterpart of free use bands is that no availability or communication quality can be guaranteed: they depend on participant fairness. No user can ask for particular formal rights or priorities due to his former activity. Nobody may complain about any newcomer who disturbs existing practices, if he complies with the regulations. It will be shown, in Chapter 10, that many efforts are made to design new radiocommunication equipment which can be used in such common bands with self-adaptive features. They make them resistant to frequency sharing conflicts.

However, today, most bands are shared between operators and services with assignment procedures to guarantee their availability and quality. As already mentioned, assignment is an authorization given to a station to use a particular frequency channel. It creates priority rights which can be opposed to any interfering installation in the future. Thus, it must be registered in order that such rights be publicly known and cannot be contested. In fact, only registered assignments create rights and can be opposed: without registration a station does not "exist". It is the responsibility of the new station operator to ask the regulator for his assignment registration. The regulator files it on a national data base or notifies the ITU Radiocommunication Bureau to be filed in the MIFR. If the assignment is relevant to a plan, this will be updated and take the new station into account. The corresponding priority rights extend to the radio links established from the station concerned. They can be enlarged to a whole network around this station.

Only a state administration or an entrusted authority may assign a frequency or a channel and be responsible for this assignment to foreign administrations. If any interference problem occurs across borders, the administrations must manage it. Frequencies assigned individually from such administrative procedures may generally be taxed according to administrative pricing methods which are part of the national regulation.

8.3. Exclusive bands: preferential sub-bands

Assignment is a pretty heavy procedure with several technical and administrative phases and formal steps which may last for weeks, even months. If it was necessary for every transmitting station of a particular network to request from an administrative bureau such a procedure, to assess a frequency channel availability and quality on each site, to check any interference problem, the burden would be excessive. It would be still heavier in the border areas due to preliminary coordination procedures. It would be nearly impossible to manage for fast changing networks such as cellular mobile systems which must permanently adapt to the traffic needs, and more generally for individual customer premises connection links.

Administrations may suggest criteria to limit the number of stations which deserve individual assignment, as power or EIRP criteria. A typical figure of one watt could be a reference under which the procedures are found too heavy for their benefit. Practically, formal assignment and registration seem to be of the utmost importance for the stations which are at the very heart of a network or whose reliability is essential to this network. Let us mention, as examples, satellites and main satellite earth stations, important broadcasting stations, notably when part of a plan, private mobile network bases, fixed service infrastructure stations, aeronautical radio stations and so on.

In other situations, a means to lighten the task is to attribute a full band to an operator who has demonstrated all professional competences and to give him the global responsibility of using the corresponding frequencies for implementing a particular application over a given area. Normally, this band is exclusive and the operator does not need to care about interferences if the network engineering is properly made. However, his freedom is limited. First of all, he must conform to all technical regulations which apply to the service and are designed to limit the possible interferences with other radio systems ("hard limits"). He must also take care of neighbors at the borders of his service area where coordination may be necessary ("soft limits"), if interference risks have not been alleviated from the beginning, for example by some bilateral agreements between foreign regulators.

As a general rule, exclusive rights, as granted by an authority, are limited by the rights which depend on another authority. The same principle applies to relative rights between services depending on different regulators. The ultimate referee, in case of a conflict inside a country, is the state government which is also in charge of finding a solution with foreign administrations to any cross-border trouble. In such cases, unless particular bilateral agreements exist between parties, only the formal registration of station assignments may be a strong basis to argue about priority rights.

This is why it is appropriate to use such an exclusive band allocation policy for regionally harmonized applications with *a priori* sharing of bands between neighboring countries in the border zones. These national partners agree to divide the harmonized band into preferential sub-bands, which may be a set of channels. These sub-bands are then allotted to each of them to be used in the border zones, according to a planning scheme which minimizes the interference probability. Thus, coordination procedures are normally avoided but may be kept as a complementary instrument for local adjustments. The national authorities may allocate parts of their preferential sub-bands to the different service operators.

Of course, every operator must respect any harmonized planning scheme and standard channel division. His only freedom is to distribute its allocated channels to different stations within his network infrastructure. In fact even this freedom is sometimes very constrained. As an example, when an operator gets only two or three channels as preferential sub-band in a border area, the actual situation management looks very much like the traditional assignment method.

In a border area, an exclusive band attribution is very difficult to manage, even impossible, when a service is not harmonized. A favorable context exists when bands, duplex span, channel division are common between participating countries. Preferably too, the technical standard should be the same for all equipment to optimize the band sharing parameters. If such harmonization is not achieved, conventional coordination and assignment procedures are probably necessary.

As explained in Chapter 7 (see section 7.2), such a combined policy of regional harmonization, national exclusive band allocation and preferential sub-band division between countries in the border areas, looks very much like a variant of plans. Broadcasting regulators seem to prefer international or regional plans whereas telecommunications regulators use both plans and exclusive band allocations, depending on the service. Notwithstanding the limitations, the allocation of exclusive bands to radiocommunication operators is becoming common practice for them. This is the case to the benefit of mobile phone network operators. In recent years, in many countries throughout the world, after very open and public procedures, companies have been chosen to create and operate mobile cellular networks. They have been commonly provided with exclusive spectrum resources, typically duplex bands, 2*10 or 2*15 MHz wide, for example, on whole national territories or regional areas, for 15 or 20 years. To obtain the corresponding licenses, very high prices may have been paid, after auctions or according to administrative decisions. These successful operations have directed the recent evolution of band allocation in Europe towards a more decentralized policy with a number of bands opened to operator activities especially for local customer services. Simultaneously the harmonization of networks and applications and the growing competition between operators in that region will contribute to enlarge the possible attribution of exclusive bands.

8.4. Assignment procedures

Coming back to assignment procedures, they proceed along a logical sequence which aims to satisfy a frequency (or orbital position) request for a particular station without disturbing any other party. An assignment should fully comply with the international Radio Regulations and any regional agreement and be respectful of former priority rights. If all conditions are satisfied, it can be granted and registered. It thus creates rights for the new entrant.

An assignment request should be formulated by an operator to his national administration or a designated regulator, with all appropriate technical data concerning the envisaged station and radio link: the geographical situation of the transmitting and receiving stations, antenna patterns, required frequency band, modulation type, EIRP, etc. They are gathered in a declaration form which depends on the service: fixed, mobile, broadcasting, satellite, etc. The ITU Radiocommunication Bureau has standardized such forms.

Assignment granting and registration create two different rights:

– a transmit authorization which is obtained if the new station has proved that it will not interfere with existing services (or already registered services; see section 8.6);

– a priority right which will protect the new station and links against any interference which may happen in the future.

The transmit authorization depends on the new transmitter characteristics. The promoter should provide data which may help all existing operators to calculate the incoming station impact on their own installations.

The priority right is related to the new network receiver characteristics. These receivers should not be disturbed in the future by any new station. For this purpose, the assignment form includes data not only relevant to the transmitter but to an associated network including one or several receivers, a receiving area or any set of link descriptions depending on this transmitter.

It may be noted that this assignment does not protect the new entrant against possible interferences due to existing stations with already registered assignments. The new station promoter has to check, from the beginning, that his project will be compatible with the radio environment, as it is. If doubts exist, an assignment may be granted but without a guarantee of non-interference.

The calculation of electromagnetic compatibility between stations (see Chapter 2, section 2.5) is now made by dedicated standard software packages. Knowing the station geographical coordinates and their technical characteristics, using propagation models and digital maps, they estimate instantly the radio field at any place around a new transmitting station and the received field from existing stations. The field intensity obtained from such calculations is statistically described and the interference probability can be evaluated. With the help of data files where all stations are registered, such software applications may be freely accessed by operators and make most procedures directly managed by users.

However, before any assignment and registration, even when these calculations prove to a promoter and his administration that the new station can be operated without disturbing other systems, an external cooperative evaluation must take place. The new station description form should be published to be examined by any party which may be concerned, having the opportunity to formulate observations and objections with appropriate arguments. This is the coordination procedure. As calculation tools are mostly standard, the results obtained by the different parties are normally equivalent but this formal consultation is necessary, during a limited time, to obtain comments and arguments. Chapter 7 (see section 7.3) gives some views on the coordination practice which may remain purely national or extend to foreign countries.

From his own analysis, a partner may answer the coordination request with different appreciations:

– refusal;

– agreement;

– agreement with observations or any intermediate comment;

– agreement without interference protection guarantee.

If no answer is obtained before the regulatory time, it is considered as positive. Based on the answers and comments, the concerned national regulator can decide to assign the requested frequency to the station. He takes the decision with full responsibility and will have to deal with any interference problem in the future, due to a wrong decision. Following assignment, registration can take place. If notified to the ITU Bureau, to be filed in the MIFR, the assignment must be backed up by all documents which indicate that the coordination procedure has been properly executed and favorable answers obtained.

Registration is the regulatory basis to argue about priority rights if any trouble occurs later. It is important to note that the reference which is registered is the assignment notification date and not the beginning of the station operation. A registered assignment can object to a future assignment, considering a possible interference, at a time when the "possibly disturbed station" does not exist. Such situations are common for satellite projects (see section 8.6). Thus, regulatory provisions should be made to limit the effectiveness of assignment registrations if the corresponding networks are not implemented and operational before a determined time.

8.5. External requirements: site constraints

Assignment, as an authorization to transmit a radio wave from a particular station, is a regulatory act which proceeds at first from spectrum regulations whose main purposes are spectrum efficiency and protection against interferences. It is only part of a global contract between an operator and his administration. An administrative authorization for spectrum utilization includes more constraints, in relation to laws other than the Radio Regulations and their national derivatives. An operator cannot use a regular frequency assignment in order to avoid complying with other external requirements. Let us here have a brief look at some complementary technical concerns about the wave usage which could be called "radio site regulations". They are part of these external requirements and are closely related to frequency assignment.

A first set of rules depend on construction and urbanism concerns. If not hidden, an antenna is very noticeable, often placed at the top of a building, sometimes on a pole or a tower: a structure which may be found ugly. Their increasing number due to mass services, such as TV or mobile phones, may be classed as visual pollution. Moreover, building such high structures and installing technical equipment in their vicinity may imply additional constraints: mast shrouds, security lights, access means, restricted areas, etc. Construction and environment regulations are increasingly addressing radio sites and are adapted to cope with these new objects.

Managing radio sites is becoming a full-time business and good places a scarce resource, with growing prices and municipal vigilance. In fact, availability of an appropriate site is a prerequisite for any new station implementation. Specialized companies have appeared which search, adapt, equip and hire radio platforms with convenient technical features for operators. For this purpose, historical telecommunications and broadcasting companies have found new market opportunities for their traditional sites. The operators, notably cellular network operators, who should manage thousands of sites, try to limit their number, utilizing common masts or platforms and multiband antennae. To find new places, antenna camouflage is often necessary for better integration of radio in towns and landscapes.

Site managers must apply good radio engineering practice for the different stations which are installed, in order to maintain their electromagnetic compatibility (see Chapter 2, section 2.4). In addition to coordination procedures which are directed at long distance interference analysis, local precautions are necessary on platforms, which are not properly covered by the Radio Regulations. Intermodulation processes have been briefly described (section 2.4). Another common difficulty occurs when high intensity and low intensity radio fields with close frequencies simultaneously exist at one place. Here, if a receiver listens to a low intensity wave, it may be saturated by a local high intensity wave because its input filter cannot eliminate such a powerful signal. The C/I ratio (see Chapter 2, section 2.3) is then so degraded that demodulation is impossible. Technical solutions can be found on a case-by-case basis such as improving the antenna relative situations, using sharper receive filters, changing frequency and so on. It has also been mentioned that national legislations may authorize restricted radio areas (see Chapter 3, section 3.7). For specific situations such as these where administrative rights and technical problems are mixed, expert administrative arbitration can be useful.

For years now, some concerns have been expressed on the social impact of radio systems, based on facts or hypotheses. Notably the media have echoed to fears concerning public health which have appeared with the growing number of mobile telephone relays. Based on calculations of the electromagnetic energy which is absorbed by human tissue when crossed by waves, with appropriate safety margins to take care of unknown biological effects, some recommendations have been issued. They fix the field intensity at a limit which a human being should be able to stand in normal conditions.

Concerning the actual risks related to common radio usage, there has been no evidence of any danger or even significant biological impact although many research programs have been conducted on these matters. Notably, all scientific reports point out the negligible contribution of base stations outside their very close radiation pattern. Generally, the field intensity, created everywhere by all fixed stations, is much lower than that created by a mobile phone placed by the user's head. Thus, if any health effect is ever proved, more care should be taken of the radio terminals than of fixed stations.

A European recommendation was published in 1999 (1999/519/CE, 12th July 1999) which provides reference values very similar to other international regulations. It requires to limiting the intensity of electromagnetic waves around human beings so that a maximum specific absorption rate (SAR) is not exceeded in any part of the body (for example 2 watts per kilogram for a human head). Such restrictions can be translated into maximum intensity of radio fields, depending on their frequency. Some values are given below.

Band	Field intensity (volt/meter)
1 to 10 MHz	$87/f^{1/2}$ (with f in MHz)
10 to 400 MHz	28
400 to 2,000 MHz	$1,375\ f^{1/2}$
above	61

It can be seen, for example, that for GSM frequencies, about 900 MHz, the recommended maximum field intensity is 41 volt/meter. It is 58 volt/meter for GSM1800.

An international measurement protocol has been published for a local field intensity description. This protocol should be strictly followed to obtain reliable data because radio fields are fast changing, notably from place to place in urban areas and according to instantaneous traffic. It is not at all easy to carry out stable and repeatable measurements. Two measurements are made successively: a broadband one which gives a value of the global field then a selective one, using a spectrum analyzer, to characterize the different contributions of neighboring stations. Some countries publish comprehensive maps of radio sites and local measurement results (for example, see the French website www.cartoradio.fr).

Operators are anxious to be fully in-line with such public directions concerning possible health effects of waves. They take care to limit the transmitted power so that it remains well under the maximum recommended field intensity value. As mentioned above, this is not really difficult a few meters away from an ordinary station antenna. In the very close vicinity, signaling posts must be placed and area access restricted, if necessary.

Another example of social impact may be found when people use their mobile phone so freely and loudly that they disturb their neighbors (theater performance), create risks (hospitals, planes) or affect public security (prisons). A simple notice to turn terminals off may not be sufficient. Thus, active jammers are sometimes thought to be a solution, covering the susceptible area and making mobiles ineffective.

Jammers are pieces of equipment which completely contradict the Radio Regulations. When they transmit a radio noise all around to overcome any mobile service, they offend against these international rules and most national provisions on spectrum matters. Their only acceptable use (apart from public security or national defense circumstances) could be inside a private place without disturbing any outside public area, which is a very difficult challenge. Some tentative regulations have been sketched but they failed to achieve clear policies.

In Europe, CEPT has made a strong recommendation (ECC/REC 04-01) to ban frequency jammers and forbid the sale of such equipment as also being contradictory to the RTTE directive on terminals. This is clearly a subject where laws may be contradictory, being based on different and incompatible social values. CEPT points out that even security objectives cannot justify jammers: on the contrary they may disturb systems dedicated to security.

The right way to ban a particular radio service from an area, which may be sometimes a pertinent objective, is first of all to include such a specification in the frequency administrative authorization, or license. It is common practice to restrict an authorization to a limited geographical area and no regulatory statement objects to excluding a particular site from the service coverage, even if it is not easily done today. In fact, such techniques could be really useful for protecting some radio sites, such as radioastronomy centers, or managing electromagnetic compatibility by sensitive equipment as radar. New tools, such as real-time radiolocation, software controlled radio, electronic maps and others may help to achieve this objective of a dynamic and precisely controlled radio coverage.

8.6. Satellite systems

The same basic regulations apply to satellite systems and terrestrial systems. However, space networks need to take more care in their assignment and coordination procedures. Satellite coverage with its international dimensions, the earth stations being spread in different countries, the occupation of an orbital position, international goods, make the whole operation to obtain various agreements very complex.

A satellite project includes one or several space platforms (satellite) and earth stations. To coordinate these different components with existing or projected systems there are many concerned parties.

The project of a new satellite or set of satellites must be introduced to the international community by a particular national administration, since only state administrations are recognized as Members by the ITU. This notifying administration should provide the necessary technical information and submit it to all other concerned administrations for examination, according to the regulatory coordination procedures. These satellite characteristics and parameters will be notified to the Radiocommunication Bureau, if the coordination is successful, to be registered. The priority rights, when obtained, are the property of the notifying administration. It is a privilege of this administration to give any compatible satellite system the benefit of such international rights. Of course, very often, the project introduced to the ITU is based on an already designed system, proposed by a promoter.

Things are different for the future earth stations which will communicate with the projected satellite. At first their coordination with other satellite networks must be examined. When installed, will they disturb any other satellite system? This study is the responsibility of the notifying administration of the new system, being considered as the designer of a global space network, including earth stations. If any interference is created later by an earth station, in normal operating conditions, this administration will have to deal with it. On the contrary, coordination of these earth stations with neighboring terrestrial stations or other earth stations should be managed by the country administration where they are installed. In any case a national administration is in charge of notifying about any earth station installed in its territory. Such provisions are related to the idea that designing a new space system, satellite and earth stations, should be done with a global view on the orbit-spectrum resource utilization, as international goods. All stations to come later must be exactly compatible with the initial assigned resource, no more. On the contrary, the "terrestrial" spectrum resources shared between terrestrial and earth stations should be normally managed by the administration where they reside, as their national goods, possibly with conventional cross-border coordination procedures.

Coordination and notification procedures are very strategic when designing a satellite project. The availability of adapted orbit-spectrum resources is a prerequisite to any development and implementation, long before a satellite launch. Taking into account the necessary time and cost attached to such big projects, spectrum affairs must be cleared as early as possible to notify their basic parameters to the Radiocommunication Bureau in order that they are registered and have priority rights granted.

This time constraint to get definite spectrum resources means clearly that "time is money". Thus, some with the help of a friend administration may speculate, publishing "paper projects" which ask for a part of the unplanned orbit-spectrum resource. If the corresponding rights are obtained, they can be sold with a profit to operators who are looking for resources. Another trick may be to prevent the development of a real system, such as a business competitor, thanks to a "paper project" in a close position which will oppose this coming system with rights granted to a phantom satellite. Such speculative actions were very popular in about 2000, at a point where the ITU Bureau was overloaded with submitted files. To oppose abuses, procedures should be improved. The simplest way is to shorten the time validity of notified network rights, depending on their real implementation agenda. Typically, the rights granted to a satellite project lapse seven years after its initial publication, if the platform is not working at that time. It can be also envisaged to make administrations pay for having their satellite rights maintained.

Chapter 9

Spectrum Monitoring

Spectrum monitoring equipment and teams should watch over frequency users to check that they are operating according to the regulations, either the general rules or their individual license. They also analyze any problem which the radio networks may face, mainly interferences: they help to find the reasons behind the problems and suggest solutions to put them to an end. They provide expert arguments for court claims, if necessary. Finally, they give a detailed view of the "radio landscape", as a description of the radio fields over a geographical area with the local spectrum usage and indications on its efficiency and quality.

Spectrum management cannot be only an administrative activity based on regulations and theories. It must take into account the actual activities and constraints of radio users, in the field, to improve the rules and procedures and make them more efficient and realistic. Spectrum monitoring provides a "view on reality" with measurements which describe the invisible radio field at any place and illustrate phenomena. It is an essential tool to guide the network and station engineering in complex situations and notably during exceptional public events.

Moreover, new prospects are opening up for operational spectrum management and frequency assignment when using automatic spectrum monitoring tools as real-time probes to describe the local radio context and find the most appropriate spectrum resource available anywhere for a particular link, depending on its radio environment. Replacing administrative band allocations and frequency assignments with such self-adaptive radio systems under the control of spectrum monitoring features may appear to be a promising idea to improve the global spectrum management efficiency.

9.1. Spectrum monitoring technical tools

It has already been mentioned (see Chapter 3, section 3.3) that radio wave propagation depends greatly on the frequency. Thus, various instruments are needed to monitor the different bands and radio systems.

Furthermore, two sets of equipment provide complementary analysis and are both necessary:

– permanent fixed supervision base stations may automatically observe wide frequency bands for long periods with the objective of assessing the spectrum global quality in their neighboring region and monitoring the radio stations within their measurement range. They can also detect any abnormal situation or event which may need inspection;

– mobile stations are operational tools which can be driven to a radio station vicinity when any problem occurs or for an inspection. They can provide detailed data on radio parameters, depending on measurements realized by technicians

9.1.1. *HF band monitoring*

HF monitoring instruments, adapted to the 3-40 MHz band, can observe either short distance radio waves using ground propagation or very long distance communications using ionosphere reflection. Among the important usages of such links, let us mention fixed services, amateurs, maritime and aeronautical communications and beacons, audio broadcasting, etc.

Figure 9.1. *HF monitoring center. In the distance, great rhombic antennae can be seen, dedicated to HF international link reception. At the front of the picture, is a part of an HF goniometer to localize distant transmit stations. The HF monitoring activity is planned within a measurement program concerted with ITU*

The receivers of a HF monitoring center automatically scan a particular frequency band and measure several characteristics of all received carriers: central frequency, modulation class, field intensity, bandwidth, and compare these data with those of registered stations in the band. In so doing, they provide indications on the spectrum occupancy and may detect any abusive station and help solve interference problems. The measurement programs are organized internationally, notably by the ITU according to Radio Regulations articles, and conducted by several monitoring centers which cooperate. They are often realized during the night when ionosphere propagation is favorable.

Due to wavelengths (decametric waves), the HF monitoring centers are pieces of large equipment, installed far from cities to benefit from a quiet radio environment. They are provided with different antenna sets: omnidirectional, fixed rhomboidal, electronically directable arrays, etc. to fit as well as possible with the measurement protocol. Every significant country has at least one HF monitoring center, but large countries, the size of a continent, may have several spread over their territory.

HF goniometers are sets of antennae regularly spread over a large area, typically a square kilometer, with a precise geometric implementation. They can measure the horizontal angular direction (azimuth) of a distant transmitting station with a precision of 2 or 3 degrees. They can also roughly evaluate the distance of this station from the ionosphere propagation characteristics at a time, the ionosphere layers acting as a mirror for waves. However, such limited performances are sufficient to describe the position of a HF transmitting station on the Earth with a precision of a few tens of kilometers at a distance of several thousands of kilometers. When several HF centers are simultaneously aiming at the same transmitting station, being in such relative geographical positions that the evaluated directions can be crossed, the distant station location is much improved. Altogether they provide the concerned administration with precise information to find any faulty transmitters. These combined performances are so attractive that European monitoring centers are studying a possible interconnection of their equipment to make such cross measurements automatic.

Under 3 MHz, long and medium waves are propagated as ground waves, at a limited distance. They can also be monitored by the same centers but the corresponding transmitting stations, mainly for broadcasting, are perfectly well known and do not need to be located.

9.1.2. *Metric and decimetric band monitoring*

VHF and UHF bands, from 30 MHz to 3 GHz, are at the very heart of the radio spectrum for short and medium distance communications. They are the most used by mass equipment and services, notably for broadcasting and mobile services: FM or DAB radio sets, TV receivers, mobile phones, RLANs, GPS and GALILEO location terminals, SRD and LPD devices, wireless telephones, all of them work in these bands. Most security services and networks: police forces, medical teams, fire brigades, aeronautical and maritime equipment, have their communications carried by such waves, as well as many defense applications.

Thus, as a strategic resource, these bands are coveted and should be carefully monitored. They are so attractive that most terminals using unassigned frequencies, which are freely introduced on the market, want to use them, sometimes without any concern for national frequency allocation tables, often on frequencies dedicated to security networks. For example, some bands for defense applications may be left unused in ordinary circumstances but must be kept available for crisis periods: thus, they may be thought "empty" and usable by unscrupulous users or manufacturers. On such a matter, the regulation of terminal markets may be thought ambiguous, notably in Europe where some of them may be sold freely but should not be used in the same country.

VHF and UHF waves are mainly propagated by line of sight, however with disturbing phenomena such as reflection, refraction, attenuation masks and so on (see Chapter 3, section 3.3). To permanently monitor spectrum use, a dense network of monitoring stations is necessary, with remote receivers spread at a distance about 100 kilometers from one another, covering all "useful areas", notably urban zones where the radio system density is the highest. The best situation occurs when any dense area is surrounded by two or more monitoring stations which may cross their measurements. The sites to install such stations must be carefully chosen to achieve an appropriate coverage. High sites are preferred to extend the radioelectric view at a maximum, however they should not be shared with powerful transmitters, such as broadcasting ones, to maintain enough measurement sensitivity.

The monitoring equipment, notably the measuring sets placed in such stations, can be more or less sophisticated. Their minimum component is a precision selective voltmeter with scanning facilities, or spectrum analyzer, which can periodically observe chosen frequency bands and file the main radio parameters of detected carriers (center frequency, intensity, frequency deviation, etc.), to compare them with registered stations, in order to point out discrepancies. The stations are fitted out with calibrated broadband antennae and goniometric sets to estimate the azimuth of transmitters. Simple demodulation devices are useful for analog modulated carriers to identify their operator when using selective call procedures.

Figure 9.2a. *VHF-UHF monitoring station. Three goniometric antenna sets can be seen. At the top: 500-1,300 MHz, 150-500 MHz, 20-160 MHz. At a lower stage are "discone" antennae which are used for measuring carriers and listen to signaling messages; one is covering the 400-1,300 MHz band, the other 100-500 MHz. It should be remembered that, for a given gain or directivity performance, the antenna size is directly related to the wavelength*

Part of the complexity and cost of a fixed monitoring network for VHF and UHF waves is related to its necessary remote automatic control. Due to their number, it would be very expensive to let operators stay at every station. Furthermore, the bands to be monitored and the signals analyzed are so important and change so often that a software program is necessary to pilot the measurements. The results should also be transmitted to a central computer for interpretation. Such a data processing system and related transmission links are very costly. They should be considered and engineered as part of an integrated spectrum management system where data files on radio stations must be connected to administrative procedures to be fully reliable and updated.

A challenge to take up for modern supervision monitoring systems is relevant to new digital modulations which are now commonly used by most customer equipment. There are so many various types that they cannot be easily recognized, evaluated and demodulated as analog ones were. Thus the radio carriers remain anonymous or even undetected, except when sophisticated signal processors can be used. Even basic carrier parameters may not be easy to measure, notably for spread spectrum signals.

On the contrary, the location of traditional VHF and UHF transmitting stations from remote monitoring centers is very efficient and precise. The uncertainty is often less than a kilometer and can be easily reduced using mobile equipment.

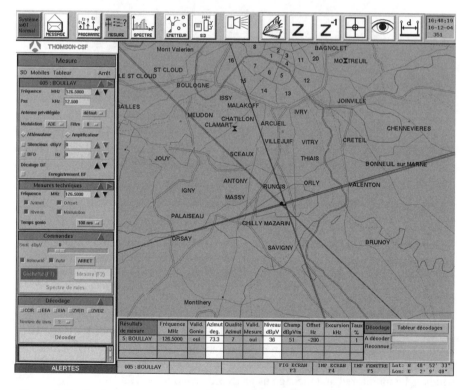

Figure 9.2b. *Goniometric measurements from three UHF monitoring stations around Paris. Their combined indications provide an excellent location of a transmitting station (here at Orly airport)*

9.1.3. *Microwave monitoring*

Waves whose frequency is higher than 3 GHz are called microwaves (SHF and EHF). As path attenuation increases as the squared frequency value, such waves can only be used at a short distance or with the help of directive antennae which provide high gains and sharp directivity.

In the first case, the microwaves cannot generally disturb any distant link, due to their fast attenuation. If any interference occurs on neighboring systems, the relevant microwave transmitters and receivers have to be located in the close vicinity from local indications and evidence.

In the second situation, waves stay contained in the straight direction between the transmitting and receiving stations. Their direct path lies well above the ground and is difficult to intercept. Only waves transmitted by antenna side lobes can be detected in the transmitting station vicinity. Thus, monitoring a microwave station from outside is not easy, notably for the fixed service. A receiving mobile high gain antenna is needed with a powerful amplifier. It should be installed at a short distance from the station and directed precisely towards it.

For such reasons, fixed monitoring stations are not usable for microwaves. Only mobile measurement laboratories which can be driven near a particular transmitting station can detect waves and perform conformance tests. Outside the close vicinity of a terrestrial microwave station, it is very difficult to intercept any useful signal with ground-based equipment. Radio laboratory vans can only be used to make limited measurements in the vicinity of well known microwave stations rather than to supervise the general microwave spectrum.

Figure 9.2c. *Mobile monitoring station for SHF waves.*
The parabolic receiving antenna can be oriented in any direction

9.1.4. *Satellite monitoring*

The telecommunications satellites are managed extremely carefully. They mainly use microwaves which are very stable carriers, in almost free space conditions. They are known by the ITU registrations with a high precision, notably their nominal orbital characteristics which a satellite must strictly comply with. Any possible interference problem has been studied from the beginning, during the coordination procedures. They can be observed from large coverage areas on the ground, without any mask. They are controlled by skilled people with powerful instruments, under the responsibility of their notifying administration.

All these conditions guarantee the full compliance of satellite systems with their nominal technical parameters. The question may thus be asked: is satellite monitoring useful? Doubts can be expressed but no control would also be inappropriate. Some countries have built satellite monitoring stations to obtain information about trouble situations, from independent expert sources. Some purposes are listed:

– routine checking of main satellite parameters: orbital location, carrier frequencies, PFD, etc.;

– providing technical data to the ITU Radiocommunication Bureau when disputes take place between satellite operators;

– measurements to back electromagnetic compatibility theoretical models;

– illegal satellite use prosecution;

– interference assessment and measurements to locate disturbing stations.

These two last objectives are the most difficult to obtain. Observing a civil satellite is easy. On the contrary, detecting an earth station which disturbs a satellite system is a difficult job. Still cleverer is locating an illegal transmitting earth station which may use unoccupied bands on a satellite for short periods. Due to their directivity, with their antenna oriented towards the satellite, transmitting earth station uplink carriers cannot be easily monitored from the ground, as with any microwave link. They can be more easily monitored from the downlink satellite carriers which are commonly a frequency translated image of uplink contents, redirected to the Earth by the satellite. This content can thus be analyzed and useful information obtained from signal processing.

It is also possible to roughly locate a transmitting earth station from differential data provided by two distant measuring stations aiming at two satellites which are receiving the same disturbing wave. The measurement protocol is critical and data interpretation needs computer assistance, for limited precision.

A satellite monitoring station should be equipped with performing antennae and receivers. Their useful frequency range may be as wide as 1-30 GHz with an accurate calibration of every frequency band. As geostationary satellites are the most common, monitoring stations may only be adapted to such almost static platforms, with limited tracking capabilities. However, if non-geostationary satellites should be observed, a station needs sophisticated mechanical characteristics adapted to fast tracking. Furthermore, as measurements may concern very weak carriers or sensitive parameters, the receiving system must include a powerful digital signal processor which can extract appropriate information from noisy messages.

Such monitoring stations are very costly and the limited task to be performed does not need many of them. It seems convenient that several countries agree to cooperate with common tools. It has been decided, in Europe, to use the German station, at Leeheim, as such. As a matter of fact, a limited number of powerful and skilled stations, installed throughout the world, can be estimated more useful than many simple ones, in every country. It would be appropriate to implement a worldwide satellite monitoring network with interconnected cooperating stations which would share their information to monitor the whole orbital space.

9.1.5. *Mobile monitoring stations*

As mentioned above, fixed monitoring stations can only carry out an overview of the spectrum use. They can detect abnormal situations such as unauthorized stations or deviations from nominal parameters and provide rough information on their location. However, to get precise data and locate an interfering transmitter exactly, mobile monitoring stations are necessary which can be driven in the field, close to the trouble source.

Their main objective is to identify precisely the disturbance origin, probably a particular transmitting station, and to check several points about it, notably its administrative and technical compliance. This information helps to understand the interference reasons and provides elements of a report in view of any legal action.

Even when no trouble is observed, mobile stations are useful for routine examinations of registered fixed stations in order to prevent any deviation from nominal conditions. Spectrum managers need several vehicles at their disposal, equipped as laboratories, with goniometers, calibrated antennae, measuring equipment such as spectrum analyzers, location terminals and so on. They should be able to gather all technical data necessary to draw up a comprehensive expert report about any local radio situation. They can be used for many purposes other than station monitoring, such as:

– radio field measurements related to public health concerns;

– service coverage plotting of fixed stations.

Requests for such measurements are rapidly increasing, notably according to the standard "health protocol".

9.1.6. *Airborne monitoring means*

It may be useful to complement terrestrial monitoring tools with airborne means. When wide areas need to be overviewed, looking for low power carrier transmitters with an unknown location, terrestrial means are not sufficient. It may be more efficient to monitor the radio spectrum from the sky or space.

Measurements from a plane or helicopter are commonly achieved for difficult and sensitive interference situations such as aeronautical frequency jamming. Spectrum monitoring can also be done with the help of low orbit satellites for military purposes. Civilian satellite versions can be envisaged. However, although interesting, these equipment and operations are very costly, too expensive when compared with common civilian needs: they can only be considered as a bonus for conventional terrestrial systems. If any project should be decided, it would certainly be an international cooperative investment.

9.2. Radio station inspections: major events

Spectrum monitoring, as described above, achieves wave overview with external tools and measurements. It may be usefully continued by station and site inspections, since waves are produced by transmitting stations installed on particular sites which are the true sources of interference troubles, if any. Thus, a periodic look at their equipment and operational conditions may prevent problems. Stations and sites must also comply with their operator license and some complementary regulations, if they exist: this should be verified.

Radio station inspections can only be selective. In ordinary conditions, it would be impossible and probably useless to periodically check all transmitting stations. However, some categories, where security concerns are imperative, should be inspected every year, such as aeronautical and maritime. Considering ordinary stations, only powerful ones need to be registered and sometimes inspected, notably when several of them share a common site. It is the responsibility of operators to build and maintain stations in conformity with the regulations and state-of-the-art standards. An objective of inspections is to put gentle pressure on them in order that they take permanent care of their obligations. Simultaneously, inspections provide operator technicians with assistance and help to improve the station engineering and explain the Radio Regulation changes. Finally they give the opportunity to update the administrative files where stations are registered. Some check points follow:

– technical radio parameters conformity, geographical coordinates, site description;

– compliance with administrative constraints, frequency coordination, assignment and registration;

– good practice for site management and maintenance, security features, mechanical structure quality;

– neighboring radioelectric field values.

Of course, such inspection programs depend on national regulations. They should be planned through the cooperation of regulators and operators in order that an appropriate sampling of stations and sites be chosen: all types of services may be concerned and all regulation aspects covered.

In special circumstances, a more comprehensive spectrum control program must be organized. This is the case when a major social event is organized somewhere, such as an international sport meeting, an historical commemoration or head-of-state conference. Thus, radio services which are installed by different operators are essential for the success of the meeting and often politically sensitive. Temporary security mobile networks, broadcasting feeder links, portable microphones and cameras and short distance communication systems must work perfectly. Furthermore, the corresponding equipment may be provided by foreign technical teams using different bands and standards which adds complexity. At the same time, the capacity of existing services, such as public radiotelephones, should be much increased to cope with the needs of visitors and attendees.

The density of radio stations and terminals and the quality of communications require perfect organization and discipline, especially when constraints are imposed on participants. As the available spectrum is limited, all the requested temporary frequencies cannot be assigned and the operators may be tempted to use channels which they imagine to be free, without taking care of possible reservations.

To prepare for such an event, from a spectrum manager's point of view, it may be useful, a few months before, to plot all the radio systems which are operating in the area concerned, file their characteristics and check that they comply with the regulations. At the same time, any unauthorized transmitting stations should be interrupted.

Simultaneously, requests for temporary frequency assignments are received from many operators and administrations. A comprehensive description of their desired radio networks should be required with detailed information concerning the number of terminals and all appropriate technical features. A major objective for successful radio global implementation is that all stations and terminals be known and referenced, with their characteristics, before entering the event area. It is useful if an identification label is stuck on any terminal. When all frequency needs have been collected, arbitrations can take place, depending on the available spectrum and priorities. Thus temporary frequency assignments can be notified to operators.

During the event, several mobile inspection teams with their laboratory vehicles should stay in the vicinity. They must supervise the spectrum and detect any abnormal use. They analyze interference or radio engineering problems and propose solutions. They enforce the technical and regulatory provisions which have been decided by the spectrum manager. Portable goniometers help them to locate any radio transmitter with the assistance of several mobile goniometers installed at the event area border. To achieve their job, they will have been provided with documents where all authorized terminals and equipment are listed, with their technical and operational parameters, notably their frequency assignment and nominal power.

9.3. Claim for interference: legal prosecutions

The regulatory provisions and all administrative procedures are directed to allocate to every user appropriate spectrum resources and to prevent interferences between systems in order to guarantee the best possible quality of radiocommunications. Thus any unauthorized emission or any carrier which does not comply with its nominal parameters may interfere with other links. Sometimes, even compliant stations may also produce unacceptable interferences due to propagation conditions (see Chapter 3, section 3.3), station engineering defaults (see Chapter 8, section 8.5) or assignment mistakes. Technical expertise or arbitration is necessary to study and clear up such situations where conflicts may occur between operators. An administration bureau should be appointed to receive complaints from users who experience harmful interferences. It will require from spectrum monitoring teams the necessary technical data to complement the information provided by the complaining operator. It will also initiate formal procedures which will bring about an orderly solution.

The same procedures exist at an international level. In every country a central bureau should be appointed to receive notifications of interference complaints from abroad. These different national bureaus have to work together, being also ITU interfaces, to manage the international interference situations. Of course the same bureau may be in charge of both national and international cases.

At first the technical and administrative procedures which follow a claim for interference are not judicial. Technical expertise is required from a spectrum regulator to study the case, give an opinion concerning the relevant rights and responsibilities and propose solutions to end the disturbance. An interference report may be drafted which summarizes the technical situation and describes the administrative context with conclusions on the rights and duties of parties. In most circumstances, the operators follow these recommendations based on expert opinion. However, in some situations, the case may be introduced before the court where the interference report is considered a major argument.

The authorities in charge of spectrum management may also go to court when they encounter illegal situations, even without any actual harmful interference. Depending on national legislations, different offences may be found, the most common being the use of private terminals with unauthorized frequencies in sensitive bands, such as aeronautical ones. Here also a detailed report is necessary to convince the judge of the administrative or technical rules which have been abused and also of possible risks or damage related to the situation.

Today most judges are not familiar with spectrum regulations which are a rather technical matter. Thus, only expert reports with clear and strong arguments can be taken into account. Hopefully a better understanding of the spectrum management rules is progressing among the professional actors and customers. Offences, when detected, may be thought more deliberate than unintentional and thus, claims can be more easily argued.

9.4. "Radio landscape" description

To increase the manager's control of the spectrum resource, improve its efficiency, resolve interference situations, and answer any question from society about radio matters, an ever better knowledge of actual radio fields is necessary. This knowledge should be more and more local to cope with operational or environmental concerns but also global, with statistical descriptions, to improve the technical regulations and design standards.

Two convergent methods are available to obtain this information: one based on administrative procedures, the other on local field measurements.

The "administrative" method uses data gathered from regulatory procedures, notably frequency assignments to stations and related site descriptions. Operators should commonly ask for authorizations when installing new radio stations, at least for powerful ones. Thus, their corresponding data can be filed and used to calculate the radio field in the vicinity with the help of computer software and digital maps. This information can be sufficient to provide useful data for particular needs:

– public inquiry on environmental matters before building a new major infrastructure, as a recent example, wind turbine farm projects are anxious not to disturb broadcasting;

– frequency coordination and assignment (see Chapter 8);

– major event spectrum planning (see section 9.2).

Furthermore, spectrum managers and regulators keep close contact with network operators who can provide more detailed and updated information on their own installations with sophisticated tools to describe their service coverage area.

However, the reliability of this administrative method is limited. When too many small transmitting stations are not included in the files, if data concerning authorized stations are not properly updated, if operators forget some procedures, the method cannot provide precise enough information, at least in some bands, after a time. This "registered" description must be periodically checked from real observations and enquiries not to become obsolete. This is an important task of spectrum monitoring to maintain the quality and reliability of administrative files.

A spectrum supervision network with fixed monitoring stations provides good information to confirm the file contents. These stations periodically observe all radio transmitting stations within their detection range. They can follow the authorized stations main characteristics on a long-term basis. The "radio landscape" as seen by a fixed monitoring station can be registered and any change detected when the scan scenario is replayed. Other interesting information that only fixed stations can provide, is a measurement of the effective load of the spectrum use. A frequency assignment and station registration does not mention any activity parameter. Thus it cannot say whether a frequency channel is permanently active or mainly in a standby position. Does it work with a load peak hour or with a stable power? Only periodical observations from a fixed point can answer such questions in view of modeling the spectrum activity to increase its efficiency.

The mobile monitoring equipment can also be used to obtain complementary information on spectrum activity description, with detailed investigations about particular bands, sites or transmitting stations. As mentioned above, at one moment a comprehensive spectrum survey can be made within a limited area, or a permanent sampling policy can be managed, day after day. The results can be compared with the file registrations to amend them accordingly. Such means are costly and time consuming. They cannot be used on a large scale, unless sensitive bands should be especially monitored. Their common purpose is pedagogic in order to maintain pressure on operators in order that they remain reactive to procedures for a global quality process. They also provide strong arguments when a debate takes place with people unfamiliar with spectrum matters, as an example for local health concerns, who are not convinced by purely administrative data. When such data can be backed up by actual measurements, made in the field, they obtain more efficiency and the validity of theoretical models is more easily recognized.

9.5. Terminals

The public accesses the radiocommunication services by using terminals which may be freely bought from the market. Most terminals are transmitters which, as any radio transmitting devices, must comply with the regulations. However, customers cannot be requested to follow any administrative procedure before using them. Thus terminals should be designed to comply with all technical or administrative constraints. Marketing should inform customers of any restriction of use.

Until the end of the 20th century, any radiocommunication terminal would have been the type certified by the administration before entering the market. This has no longer been the case in Europe since 1999, when a European Directive, R&TTE (Radio and Telecommunications Terminal Equipment), referenced 1999/5/CE, enforced new rules.

Different situations are considered depending whether the relevant service is harmonized or not throughout the European Union. If a band is harmonized, the corresponding terminals can be freely used in all countries. If not, usage restrictions may be indicated for different countries. In any case, manufacturers are responsible for their products and guarantee their compliance with regulations.

On any terminal, a quality mark, CE, must indicate that this product complies with "essential requirements" such as consumer safety, spectrum efficiency, electromagnetic compatibility, etc. An expert committee of the European Commission overviews the implementation of the R&TTE Directive in the Union. National regulators are in charge of enforcing it locally.

Part 2

Managers and their Practices

Chapter 10

New Technical Perspectives and Impact on Spectrum Management

For some time, the traditional frequency channel assignment procedures have been criticized for being too slow and rigid. Allocating a specific band to every service or application, thus attributing exclusive channels to operators with particular rights, duties and specifications lead the spectrum to become a patchwork made of smaller and smaller incoherent pieces, more and more difficult to adjust and move. This situation is partly inherited from history and related to the first radio technologies, notably to analog modulations which require a high protection against interferences and are bandwidth limited. Furthermore, until now, digital modulations have adopted similar frequency schemes and management practices, often being obliged to use the spectrum resource as it was already shared and to cope with the procedures as they were regulated. New perspectives are needed to access the spectrum more easily.

Today, promising radio technologies are currently being developed and implemented to provide enhanced wireless fixed transmission, nomadic, mobile and broadcast services. These developments have a potential for disruptive impact on spectrum use and management. Some of them will be described hereafter. Their survey has been performed within the SPORTVIEWS Project (spectrum policies and radio technologies viable in emerging wireless societies, www.sportviews.org) whose report has been partly used. It is not limited to a specific part of the radio spectrum, although the focus will be on frequencies below 5 GHz, which is considered the prime frequency range for fixed, mobile and broadcast services to customers.

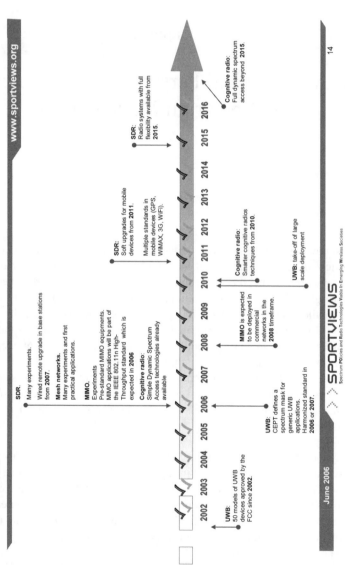

Figure 10.1. *Technology roadmap estimation*

10.1. Spread spectrum technologies

Traditional spread spectrum technology is based on the concept that the narrowband message or modulated carrier is manipulated (scrambled) prior to transmission in such a way that its profile in the frequency domain changes significantly: the radio signal occupies a much larger part of the spectrum, either instantaneously or over a certain time, but with a lower power spectral density. This processing requires a pseudo-random code which is, in the original concept, only known to the parties at each end of the radio link. Spread spectrum technology was invented in the 1940s and has been used extensively since for military and other applications that require robustness and resistance to jamming or eavesdropping.

Nowadays, 3G mobile communication systems use spread spectrum techniques, mainly to improve system efficiency and flexibility within licensed bands and to implement code division multiple access (CDMA; see Chapter 2, section 2.4), but the technique is even more powerful when used in unlicensed bands to relax the interference constraints between radio systems which share a common frequency band. Ultra wideband (UWB) can also be regarded as a spread spectrum technique and will be discussed in some detail hereafter, showing some interesting generic properties of spread spectrum techniques.

10.2. OFDM and MIMO

OFDM stands for Orthogonal Frequency Division Multiplexing. The principle was patented in 1966 (Chang, Bell Labs) and its practical use emerged in the 1980s with the rise of mobile communication technology. Nowadays it is often the modulation type chosen for mobile and short-range wireless (WLAN) communications, digital audio or video broadcasting and fixed wireless access (particularly for non-line-of-sight links). OFDM deserves to be considered here for two reasons. Firstly, OFDM is a powerful modulation to be combined with other technologies to improve radio performances on difficult propagation paths. Secondly and more importantly, there are some interesting spectrum management aspects related to the use of adaptive OFDM. Adaptive OFDM allows for throughput optimization and clever multiple-access schemes in time-varying channel conditions. Thus, OFDM is very often used for wireless communication systems supporting mobility and facing difficult propagation conditions (built-up areas).

With OFDM, channels can be switched off deliberately, for example for spectrum management reasons. Thus, OFDM allows the designer to shape the spectral profile of the signal (spectral sculpting) to adapt it as well as possible to the radio environment. This makes it suitable as a possible spectrum overlay technique to be used in larger portions of the RF spectrum.

As an example, for COFDM application to digital TV, see Chapter 2, section 2.4.

There are also antenna technologies that can be categorized under the term "advanced antenna technologies". The basic aim of deploying advanced antennae is to increase coverage or increase capacity by limiting the interference. To achieve this you either deploy antenna techniques on the transmitter side to direct the transmitted energy in a narrow beam towards the user, or on the reception side, to intelligently combine weak signals received by different antennae.

Advanced antennae can be applied at the network side (base station) or user side (terminal). Use of advanced antennae is often limited on the user side because of terminal size limitations and the nature of use of a mobile terminal. On the other hand, on the network side there are more opportunities for implementing advanced antennae, though the use here is also limited by space limitations imposed by site owners or zoning laws. In addition, the performance of advanced antenna technologies is dependent on the constantly changing propagation environment where mobile systems are used.

The approach for advanced antennae is to get a better signal by taking advantage of the dimension:

– space; by combining signals received by antennae placed apart;

– time; by combining copies of signals received within a specific time frame;

– the combination of time and space.

Antenna diversity techniques exploit the dimension space, while Multiple Input Multiple Output (MIMO) techniques exploit the dimension space and time. In theory, large capacity gain can be achieved with advanced antennae.

The most commonly used antenna diversity technique is space diversity. A mobile network is made up of many base stations to provide coverage. In UMTS (CDMA systems in general) as opposed to GSM the fact that a mobile can receive signals from different base stations is used to enhance the received signal. This aspect is also used on the network side, where so-called "maximum ratio combining" is used to select the signal of either of the base stations that receive the best signal from the mobile.

Traditionally all mobile operators exploit the space aspect of antenna diversity to improve spectrum use. In areas of high capacity demand, a high power base station is often replaced by, for example, five lower power base stations to offer the same coverage but to improve frequency reuse.

Indirectly, requirements imposed by regulators on maximum transmit power of, for example, CT2, DECT and WiFi, mandates the use of many small antennae and thus assures efficient use of spectrum.

The expectations of achievable capacity gain through the deployment of advanced antenna systems are high, but that was also the case 20 years ago. The main difference between the theoretical and practical use of advanced antennae is that in practice there is seldom a line of sight (LOS) as depicted in the figure below. Antenna beams are often reflected or blocked by objects (buildings, trucks, trees, etc.) leading to sudden loss of signal, causing dropped calls. This, combined with the relatively high costs of an advanced antenna has resulted in limited use of advanced antennae in cellular (GSM/UMTS) systems.

On a limited scale there are advanced antenna implementations (MIMO) for the WLAN type of systems in use today. It is possible to buy a WLAN base station in the shops today with multiple antennae, but at a premium compared to the more common single antenna WLAN base stations.

10.3. Ultra wideband

Ultra wideband (UWB) is a wireless technology developed to transfer large amounts of data over short distances, typically less than ten meters. Unlike other wireless systems, which use the spectrum in discrete narrow frequency bands, UWB operates by transmitting signals over wide portions of the spectrum (up to several GHz). For example, the FCC has defined a radio system to be a UWB system if its spectrum occupies a bandwidth greater than 20% of the central frequency or an absolute bandwidth greater than 500 MHz.

UWB was created several decades ago. However, it was only in the late 1990s that technology was available to use it in consumer electronics. After the FCC approved UWB in 2002, many international and national forums dealing with spectrum management concentrated their efforts on this modulation in the frequency bands specified by the FCC. However, it was difficult to reach definitive technical assessments concerning its merits in the absence of UWB mass deployments and experimental evidence.

UWB has a variety of possible applications. Those that are estimated to bring most economic benefits to consumers are likely to be in the personnel area networks which include homes and offices and high speed networking. Until recently, almost all data connections between electronic devices in such environments used cables (both wire and fiber optic). However, there is now an increasing interest in replacing cable connections with "wireless" links. Prominent wireless technologies deployed to date include Bluetooth and the 802.11 series of wireless LAN (WLAN) technologies. UWB is a potential alternative to these local area wireless technologies. The main advantage of UWB over existing wireless alternatives is that it should offer much faster data transfer rates (100 Mbit/s up to 1Gbit/s) over short distances, using the spectrum in a stealthy nearly imperceptible) fashion.

The main characteristics of UWB and other short-range wireless standards are presented in the table below.

Technology	Data rate	Range	Cost	Power	Spectrum	Issues
UWB	50-100 Mbps	150 m	Low	Low	3.1-10.7 GHz	High data rate for short range only
Bluetooth	0.8-1.0 Mbps	10 m	Low	Low	2.4 GHz	Speed and interference issues
802.11a	54 Mbps	30 m	High	High	5 GHz	High power consumption, high costs, bulky chipset
802.11b	11 Mbps	100 m	Medium	Medium	2.4 GHz	Speed and signal strength issues for more range
802.11g	54 Mbps	30 m	High	High	2.4 GHz	Connectivity and range problems. High cost
HIPERLAN	25 Mbps	30 m	High	High	2.4 GHz	Only European standard. High cost
Home RF	11 Mbps	50 m	Medium	Medium	2.4 GHz	Speed issues
Zigbee	0.02-0.2 Mbps	10 m	Low	Low	2.4 GHz	Standard still under consideration, very low communication range, low data-rate

Figure 10.2. *Comparison between UWB and other short-range wireless standards*

One of the important issues of UWB communications is the protection of incumbent and future spectrum users. Due to their very wide frequency band and limited power, UWB systems create an extremely low power spectral density which should not be harmful to any other system in the vicinity. They are not planned to operate under any specific allocation but there may be lots of existing and planned wireless systems operating under allocated bands within the UWB signal band. Thus if UWB can be deployed without undue interference to other allocated services, it will effectively increase the availability of the spectrum by a more efficient use of already allocated spectrum.

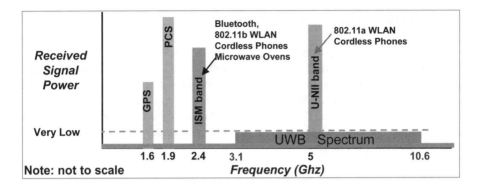

Figure 10.3. *How UWB uses the already allocated spectrum. Source: INTEL*

However, the debate is open. Some recent theoretical and experimental investigations show that UWB emissions do not cause significant interference to other devices operating in the vicinity. Typically, their low power spectral density causes UWB signals to lie below the unintentional emitter noise limits defined by the FCC regulations. As a consequence, a UWB system, as standardized by FCC in the USA, is deemed not to cause any more interference to a narrowband receiver than the spurious emissions from a computer, a microwave oven or a car's ignition. Thus, in the USA, some companies are working to deploy UWB technology and put in place an Ultra Wideband Working Group (UWBWG) to negotiate limit adjustments with the FCC. Other companies and organizations, not being convinced, are lobbying for the protection of existing narrow band allocated services against possible interference generated by UWB systems notably in their "out of band" field.

The European approach is somewhat more cautious than that of the US. Although precise regulations for UWB technology are still subject to debate within CEPT, it is expected that UWB will be admitted in Europe but CEPT may retain a more conservative approach towards UWB emissions than the FCC. The main question surrounding the introduction of UWB systems is the question of interference caused to existing spectrum users. Mobile operators generally warn regulatory bodies about the potential consequences of introduction of UWB in Europe and CEPT is carrying out studies in order to precisely evaluate such potential interferences. UWB will probably be limited to short range systems in the framework of unlicensed systems in "commons" type frequency bands and cannot be seen as a potential candidate for mobile applications offered by cellular systems.

10.4. Dynamic spectrum access technologies

"Dynamic spectrum access" (DSA) is any form of spectrum use for which the set of transmission parameters is not fixed beforehand but can be chosen and changed dynamically, depending on the radio environment. This involves the selection of the appropriate band, channel, bandwidth, transmission power, modulation, coding scheme and access method. The value of this adaptability is to automatically operate and maintain the system and network performances under different and changing radio environment conditions.

Technologies that allow flexibility in the transmitted waves exist today and are being continuously improved and extended. Some forms of simple flexible spectrum usage exist today but ongoing developments in software-defined radio technologies (SDR) will introduce more advanced forms of DSA. A key research topic is the concept of smart or cognitive radio (CR) systems. These systems must be able to sense and interpret their spectral environment, make decisions about how to adapt their own spectral behavior and evaluate the effects of their decisions (self-learning).

Automatic channel selection mechanisms can be seen as early forms of dynamic spectrum access. Several examples can be given:

– an automatic frequency selection principle can be found in modern car radio sets. The radio set scans for another channel carrying the radio station the user had selected if the received signal on which the receiver is tuned deteriorates. It is important to note that this mechanism applies to a spectrum passive device (the car receiver itself does not transmit). Another development is FM transmitters for MP3-players. They produce a low power FM signal that the car radio can tune to. The transmitter device looks for an unused FM frequency and transmits at a very low power to prevent interference with licensed radio stations. Such a basic example shows that a flexible spectrum access may be based on transmit and receive automatic features, as well;

– DECT cordless communication systems for use in residential and business environments utilize a dynamic channel allocation (DCA) mechanism. At call-setup, DCA looks for a vacant channel within 10 available in the 1,880-1,900 MHz band. The vacancy of a channel is determined through channel measurements. It is continuously monitored during a call and if the quality of the channel deteriorates below a certain threshold, the call is transferred to another available channel;

– dynamic frequency selection is a (mandatory) feature incorporated in IEEE 802.11h compliant WLAN devices operating in the 5 GHz frequency band where they have secondary status. The DFS mechanism avoids radio channels already in use by primary users (e.g. radar systems) or by other WLAN systems. Combined with a transmitter power control feature, it also improves the spreading of occupied

WLAN channels which keeps the aggregate power spectral density below limits required to protect satellite earth observation services with primary status in this band as well. DFS is performed by the WLAN terminals under control of the access point and used to make decisions if a channel change is required and what will be the best channel to jump to. This already clever DFS system points out typical difficulties to be overcome: here a specific challenge is the ability to reliably detect radar signals, especially frequency hopping radar.

The previously listed examples have common features:

– the system behavior is based on predefined algorithms and thresholds and is therefore predictable and reproducible;

– they are based on a unilateral coordination. Decisions and actions of radio systems are based on information provided by themselves, not using anyone from outside systems with peer or higher status that may coexist in the band.

10.5. Software-defined radio

The essence of a software-defined radio (SDR) is that the functionalities of the radio physical transmission level (physical layer) are almost completely implemented in software. This is a major technological change from traditional radios which operate with a predetermined built-in waveform and cannot subsequently be modified. In a fully SDR, software creates the possibility for the radio to generate a wide variety of waveforms and associated settings. The radio architecture is layered as modern computer protocols where the bottom layer includes a digital signal processing hardware and the top level, applications to parameter waveforms.

SDR technology is a logical step in the evolution of wireless systems with a number of benefits for terminals such as cost reduction, longer lifetime and multi-use capability. For spectrum management progress, SDR adaptability is also a key. Through its software, an SDR can operate on multiple channels and communication standards, using multiple modulation schemes and access methods. Thus, it is able to adapt to its spectral environment. Also, the radio can minimize the power usage and modulation bandwidth of the transmitter.

The sought benefits as well as the SDR design choices depend on the application. In the 3G/4G mobile phone domain, there is an interest in network re-configurability but re-configurability of customer mobile terminals has a lower priority. On the contrary, the military focuses on waveform portability to resolve interoperability problems with legacy systems. Both the military and the 3G/4G industry could play a role in SDR developments interesting the private mobile sector

(PMR) and especially the public security sector whose available spectrum is fragmented into small segments that reflect various improvements and frequency strategies over the past five decades. As a result, many different wireless standards exist in Europe to be used in each of these frequency bands, with various capabilities, technologies, modulations and protocols which cannot interoperate since they were never intended to work together. New digital wireless standards, such as TETRA in Europe and APCO P25 in the USA, are being developed to standardize PMR wireless interfaces and network devices, and compliance to these standards will improve interoperability. However, in a short and medium time frame the generalized system incompatibility will continue to be a common challenge. This may provide an opportunity for SDR technology.

In such a context the advantages of software-defined radio are straightforward. SDR systems can be developed to accommodate different types of waveforms and protocols, especially existing ones that are currently used in legacy systems operating in specific bands under appropriate regulations. In addition, new waveforms can be developed which are likely to have a flexible or even agile definition, depending upon the application domain, with quite different spectral properties compared to legacy systems.

From a system point of view, the spectral profile of an SDR node or network may adapt very dynamically, depending upon service requirements imposed upon the radio, local environment conditions (propagation) and possibly pricing aspects. Such chameleon behavior can be very beneficial to the performance of the node and the network.

SDR implementation poses challenges for radio regulation, standardization and global spectrum planning. As an example, spectral masks specifying frequency-dependent power density limits are probably the only viable way to regulate bands in which SDR technology will be deployed, which is a completely new method. On the other hand, the flexibility that SDR provides can also be used by the regulators to apply specific location and time-dependent regulations in some bands.

Another important issue in a regulatory perspective is the certification of SDR-based systems. Radio systems have to comply with industry regulations and spectrum regulations. With SDR, given their swift and easy re-programmability, certification and assurance concerning the system behavior will become far from trivial. Any uncontrolled software upgrade has the potential to modify radio behavior in such a way that compliance with standards and regulations may be lost. Such changes may be illegitimate (hackers) or unintentional (bugs). A risk is that the signal coming from a modified SDR interferes with signals transmitted by other users. The problem would be exacerbated if the illegitimate software modification were to be implemented into multiple radios simultaneously, thereby extending the region of harmful interference.

Views on this subject are expressed by the FCC in the rule change of March 11, 2005 [REF _Ref134949860 \n \h]. The FCC requests that SDR manufacturers take steps to ensure that software that can be loaded into the radio "must not allow the user to operate the transmitter with operating frequencies, output power, modulation types or other radio frequency parameters outside those that were approved". Also, manufacturers must provide "a high level operational description or flow diagram of the software that controls the radio frequency operating parameters".

10.6. Cognitive radio

CR is a radio that is capable of cognitive behavior, commonly described as a six-phase cycle: "Observe, Orient, Plan, Learn, Decide, Act".

The process begins with the observation or awareness phase during which the radio acquires information by itself and recognizes its environment. For example, the radio gets information regarding its physical situation including time (temporal context), space (geographical context) and frequency (physical interface context). Using this information, a fully flexible CR is able to process its perceptions and orient itself, making choices based on this existing knowledge. Furthermore, this knowledge provides a foundation for the development of alternative options in the planning phase of the cognition cycle. In a fully flexible cognitive radio, the radio will also take into account its past actions and experience, and incorporate them during the next two steps in the cognition cycle, the decision and act process.

In summary, CR is a radio that is aware of its environment (with more or less characteristics such as vacant frequencies but why not also user preferences, prevailing spectrum rules and operator tariffs), and employs this acquired information in a reasoning process that leads it to decide on its transmission behavior. Additionally, it is capable of learning through an evaluation of its own behavior and experience. An important consequence of such a cognition cycle is that the radio behavior may become unpredictable.

A radio system cognitive behavior is not strictly limited to its spectrum strategy; it may benefit from many other communication functions. However, given our focus on spectrum policy, we will primarily address the consequences of cognition from a radio spectrum point of view. Of course, software-defined radio is widely regarded as an important enabler for CR.

In the evolution towards CR, numerous technical challenges still remain. Related to the spectral environment sensing, some issues are the following:

– wideband sensing. The radio must be able to assess the actual spectrum use over a wide tuning range. A challenge is to combine a large instantaneous bandwidth and sufficient measurement accuracy. A vast amount of data must also be pre-processed to perform frequency opportunity identification;

– opportunity identification. The radio must be able to detect whether a frequency channel is in use. Several methods are available for this purpose, all of them with a delicate trade-off in the signal detection process: higher precision in a wider monitored frequency band requires more processing time and a higher power supply;

– interference prevention. If a frequency band is found to be free and therefore available to an opportunistic spectrum user, interference with a primary user is still possible. The CR might not detect any primary user signals but it does not know whether its own transmission is detectable on the site of any primary user. For this purpose it would need location information of all neighboring primary users, either provided by the primary network itself or through a costly network of sounders that monitor the spectrum;

– software challenges of SDR exist for CR as well. In addition, the decision and learning processes are still under study and may increase the risk.

The intelligence offered by CR and the flexibility provided by SDR have important consequences for the spectrum use. Already experienced features which can be incorporated for more efficiency include:

– frequency agility: the ability of a radio to change its operating frequency to optimize spectrum use under particular conditions;

– dynamic frequency selection (DFS): the ability to sense signals from other nearby transmitters in an effort to choose an optimum operating environment;

– adaptive modulation: the ability to modify transmission characteristics and waveforms, thereby exploiting transmission opportunities in the spectrum and adapting to the throughput required by the user;

– transmit power control: to allow transmission at full power limits when necessary but constrain the transmitter power to a lower level to allow greater sharing of spectrum when higher power operation is not necessary;

– location awareness combined with a registered transmission policy: the ability for a device to determine its location and the location of other transmitters. It may determine whether it is permissible to transmit and thus select the appropriate operating parameters such as the power and frequency allowed at its location.

More sophisticated features can be envisaged such as a negotiated use: CRs may eventually enable neighboring systems to negotiate for spectrum use on a real-time basis, without the need for prior agreements between parties.

Some of these features may be delivered by less revolutionary radios. However, CR use them at a higher level where there is less need for the environment to be predefined (i.e. what sort of signal, which is not to be disrupted, is expected? which radio infrastructures exist?, etc.)

Spectrum regulation must find a way to deal with these new flexibility opportunities, making room for innovation while protecting users from harmful interference. Today most national spectrum allocations are specific for a type of service, a technology or a licensed spectrum user. The new radio technologies CR and SDR ask for a more flexible approach, technology and service neutral, and may allow the spectrum to be shared by multiple users. The potential of these technologies has given rise to the sweeping perspective of "no-regulation-at-all". Whether such a situation will ever emerge depends on many factors:

– technical feasibility (are technical challenges met and can developments lead to affordable technologies that can really provide this extremely flexible spectrum access?);

– are more fundamental issues addressed properly such as external interference prevention in an unpredictable behavior context?;

– willingness of operators and manufacturers (do they want to share information on their systems and networks with outside devices? Will they agree to transfer the radio control from their "own hands" into the "terminal hands"?);

– willingness of military, public security services (to share exclusive bands).

Current solutions are limited to short-range systems like WLAN. They cannot justify alone very sophisticated and costly innovations but they pave the way.

Given these open questions we may conclude that a regulation free situation is still a remote prospect, if ever feasible at all. CR and SDR technologies will have to demonstrate that they can deliver on their promises of extremely flexible yet reliable and non-interfering spectrum use. They must do so in the laboratory but, to gain goodwill, practical applications are vital. The unlicensed bands provide an opportunity where these advanced sharing technologies may prove themselves. Successful applications in these bands may thus persuade incumbent spectrum users to allow these technologies into their domain.

On one hand, the flexibility offered by these new technologies poses a challenge to the spectrum regulator: a significant amount of control over the spectrum is lost while the number of parties using the spectrum increases drastically. Simultaneously, the behavior of systems may become unpredictable. On the other hand, an increased flexibility may induce a much more efficient use of the spectrum. The emerging CR/SDR technology may deliver flexible forms of spectrum sharing, allow negotiation of frequencies between users, facilitate access for unlicensed users when the spectrum is not crowded and overcome incompatibilities among existing communication services. Is not the benefit higher than the risk?

10.7. Intersystem control

Multiple radios making use of the same frequency band require some form of coordination. Within a single radio system this is relatively easy to implement. For instance, in the GSM system, there is a pilot channel through which terminals call to access the spectrum and the network allocates communication channels. Coordination between different radio systems is a more challenging task since a global "administrator" is not present.

As radio designs and systems gain intelligence, it is becoming possible to improve the spectrum efficiency through spectrum sharing between systems. Therefore, intersystem coordination becomes ever more relevant. The coordination process may involve the negotiation of many parameters: technical (frequency, location, time, transmit power, modulation, etc.), financial (price, payment options, etc.) and service quality (interference protection, signal-to-noise ratio, etc.).

Various forms of network coordination have been suggested in recent years, either provided by some central controller or database or performed by the radio systems themselves, without such a central entity.

Central coordination

A method to improve spectrum sharing, now under study, is based on a dedicated entity placed in hierarchy above the radio access networks. This central controller is in charge of the spectrum management and controls the assignment of resources to different radio networks. This concept of a central controller may insure flexible spectrum sharing with frequency allocations changing over time. It may also act as an information pooling system. In this case it only receives spectrum utilization updates, processes the information and provides it to the cooperating radio systems. Thus, autonomous radio systems and networks can base their own strategy from this information to choose available spectrum resources and assign frequencies for their specific communication purposes.

Distributed coordination

In this type of coordination, radio systems work together to shape some sort of coordination agreement, in order to prevent interferences. The systems may have an equal status or one of them may control the others. In the latter case the secondary systems must coordinate their spectrum use with the master. In case of equal status, both users are responsible. Two forms of distributed coordination are possible: unilateral or mutual. For unilateral coordination a transmitter has to identify a vacant piece of spectrum before it can start transmission, while in mutual coordination radio users communicate to each other their spectrum needs in order to coordinate the spectrum access. The most sophisticated forms of mutual coordination require a language for networks to communicate, a coexistence protocol. Several working groups are taking steps to develop such protocols.

The technology to provide such solutions is not implemented yet. It must still prove its efficiency, especially for the coordination between systems of different technology. In addition, legacy systems operating in the same band cannot be included in the coordination process, thus intersystem operation is limited to regions in the frequency-time-space domain without incumbent users. The following challenges for implementing intersystem control are identified:

– finding fair allocation policies, rules or mechanisms for spectrum sharing;

– how to determine a priority between systems? How to determine whether a user has the right to access the spectrum?;

– in distributed coordination, updating the policies when the systems are already in use is not straightforward as the frequency assignment mechanism is distributed between them.

Central coordination provides a way for the regulator to maintain some control over the spectrum access, even when there are multiple licensees of varying status and a wide range of transmission equipment. In fact, such a central controller may look like a computerized real-time regulator. Thus, through implementing central coordination, the regulator can safeguard against interfering spectrum use. But distributed coordination has other advantages. Not only does it provide a more lightweight form of spectrum management, it also stimulates innovation incentive to coordinate spectrum access.

10.8. Mesh networks

Mesh networks are radio communication networks in which radio nodes provide retransmission capabilities to neighboring nodes, allowing end-to-end connectivity in the network based on multi-hop routes. In structured mesh networks, the radio nodes have fixed positions. In "ad-hoc mesh networks" the radio nodes act as mobile terminals.

Structured mesh networks

The term "mesh network" (in the meaning of structured mesh network) is often used in relation to fixed wireless access (FWA) or wireless local loop (WLL) systems which provide access services to small companies, offices and residential users. Such systems are based on a point-to-multipoint architecture, in which a base station serves several fixed terminal stations within its service area. Installing multiple base stations, a larger area can be covered in a similar way as it is done with cellular networks for mobile communications (see Chapter 2, section 2.4 and Chapter 5, section 5.1). Structured mesh networking is an extension of the FWA concept where terminals can directly communicate together, i.e. ignoring the base station. The terminals are also able to relay their transmissions, so a meshed communication network is formed as schematically shown in Figure 10.4.

Figure 10.4. *Schematic representation of a mesh network for fixed wireless access (Nortel Networks)*

Ad-hoc mesh networks

Ad-hoc mesh networking is based on mobile terminals which form an autonomous network as they come into each other's vicinity. It pursues a high level of self-management and self-healing functionalities with minimal requirements for intervention of the users or a network operator. Within the ad-hoc networking framework, several network management protocols are defined e.g. protocols to admit additional terminals which come in range of the other network nodes or protocols to optimize transmission power levels and update route selection algorithms to improve the overall transmission efficiency. An example of ad-hoc mesh networking is schematically represented in Figure 10.5.

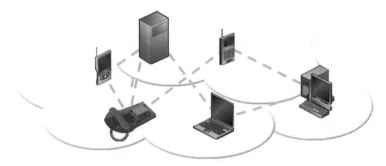

Figure 10.5. *Schematic representation of an ad-hoc mesh network (Nortel Networks)*

Their self-organizing and automatic network management features make ad-hoc networks easy to use and convenient in many applications, notably personal area networks, temporary local networks or machine-to-machine mobile communications.

Mesh networking can offer benefits compared to the more common cellular network topologies based on the deployment of base stations and central network management. Some of them follow:

– no single point of failure;

– robustness due to alternative routing possibilities in the network;

– range extension and coverage enhancement.

An interesting aspect of mesh networks concerning radio engineering and spectrum use is their capacity to overcome some propagation problems. Using higher and higher frequencies to connect terminals and customer premises with antennae placed nearer the ground make waves meet more obstacles in the radio path, such as buildings and trees, resulting in a significant attenuation of the radio signals. However, many broadband radio systems require a (nearly) free line of sight between base station and terminal (in a cellular network) or between individual terminals in a mesh network in order to establish reliable communication links. In a cellular network, it often occurs that many geographical locations that fall within the theoretical range of a base station cannot be served due to signal blocking caused by obstructions in the radio path. With the relay function in mesh networks, the coverage can be significantly improved as the radio signal can be diverted around obstacles, being relayed from terminal to terminal. This coverage improvement is considered to be a significant benefit of mesh networking, which applies under the condition that there are enough terminals within the mesh network.

Other benefits are specifically related to ad-hoc mesh networks:

– no fixed or installed infrastructure;

– networks are self-organizing;

– no need for network planning.

Of course, many challenges need to be overcome to make mesh networks broadly used, notably ad-hoc mesh networks, but nowadays research is aimed at solving these challenges and optimizing the communication possibilities offered by these networks.

It is often claimed that mesh networks are significantly more spectrum efficient than conventional cellular network topologies. The argument behind this claim is as follows. In mesh networks, every terminal is a node in the network with a relaying function for information transmission. With an increasing node density, the distances between nodes become smaller and thus less transmit power is required for the radio transmission links. Due to this mean lower transmit power, the overall interference level is lower and therefore more links are possible.

However, it is questioned whether ad-hoc mesh networks do have good scaling properties. As node density and geographical size of a mesh network increase, the information rate available to any particular user decreases. In larger networks, much of the available radio link capacity is used to relay the data transmission from source to destination along a path of intermediate nodes. This relaying completely undoes the advantages. Pure ad-hoc mesh networks do not scale: the capacity does not increase with growing number of nodes in the network. The assumption that in a mesh network users can "self-generate" transmission capacity is not valid. A solution can be found with clusters of mesh nodes directly interconnected through "cluster heads". These special mesh nodes, referred to as "cluster heads" or "relaying nodes", are more or less equally distributed across the mesh network; see Figure 10.6. By forming clusters of mesh nodes (in localized regions) and routing data traffic between different clusters on a separate hierarchical level, the scaling possibilities of mesh networks can be improved.

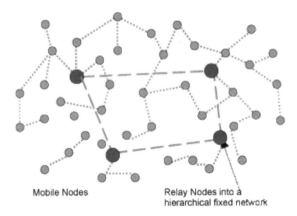

Mobile Nodes   Relay Nodes into a
                                hierarchical fixed network

Figure 10.6. *Hierarchical mesh network topology [Ofcom (UK), "Mesh networks",*
http://www.Ofcom.org.uk/research/technology/overview/emer_tech/mesh/]

Mesh and ad-hoc networks offer new interesting possibilities for communication applications. Concerning their impact on spectrum management, some observe that (ad-hoc) mesh networking is often related to autonomous operation with no or very little operator intervention. Therefore, (ad-hoc) mesh networks could be easily introduced on a license free basis. However, it can also be thought that they are only technological variants of wireless local loop or mobile cellular networks and that the same traditional regulatory practices should apply.

In fact, mesh networking is a specific utilization of low power devices and CR technology. It does not bring along additional consequences related to spectrum management. However, a virtue of such networks is to distribute very equally the power spectral density on a wide area, without any notable local peak. Their radio field may look like a "white noise" which can be easily modeled as an "interference noise temperature" and managed as thermal noise when considering possible interferences with other systems. Thus they may avoid point to point coordination or individual assignment procedures. More generally they follow the "principle" proposed for the best spectrum efficiency at the end of Chapter 6, asking for a perfectly smoothed mean spectrum distribution.

Chapter 11

The International Telecommunication
Union (ITU)

Right from the beginning of modern telecommunications (which can be dated from 24th May 1844 when Samuel Morse sent his first public message over a telegraph line between Washington and Baltimore), the need for international cooperation in order to implement these such promising techniques coherently was felt. Indeed, telecommunications systems have an obvious vocation to link correspondents living in different countries. In addition, the national governments committed to the quality and progress of telecommunications on their own territory were naturally inclined to adopt technical and administrative agreements allowing us to enjoy in the most effective way the possible common advantages brought by the new means of exchange placed at their disposal by science. For a history of the ITU see www.itu.int/aboutitu/overview/history.html.

11.1. The ITU today

Perhaps this short historical background will shed some light on the organization and the working methods of the ITU as we know them at the beginning of the 21st century. Recent evolutions, and in particular the opening to competition of telecommunications since the 1980s, has deeply modified the landscape, depriving for example the individual States of most of their responsibilities with regard to standardization. In 1992, a plenipotentiary conference, known as the Additional Plenipotentiary Conference, took place in Geneva and drastically remodeled ITU with the aim of giving the Union greater flexibility to adapt to today's increasingly complex, interactive and competitive environment. From 1st January 1993 a new ITU constitution was adopted which approximately 200 countries currently approve.

The Constitution is the main document that organizes the functioning of the ITU, specifies the missions of the organization and points out the rights and duties of Member States. It deals more particularly with radiocommunications in a certain number of articles. In this respect the "core values" that the Constitution proposes are as follows:

– the States have the sovereign right to regulate telecommunications on their territory. They are the actors of the international coordination which comes from the Rules and Regulations and organizes their reciprocal rights and duties so that the freedom of the one stops where the freedom of the others starts;

– this sovereign right of Member States follows an "ethics" code: the use of the spectrum must be rational, effective and equitable;

– the spectrum and orbits are considered limited resources, which must be used efficiently;

– the rights of protection of radio signals from interferences can arise only from the formal processes of allocation, planning and assignment of frequencies or orbital positions, in conformity with a license. These rights must be recorded. The ITU records the rights in order to be recognized at an international level;

– no one is allowed to scramble or create harmful interferences to a pre-existing station whose rights are recorded and which functions in accordance with the rules.

Three "sectors" share the basic work of the ITU:

– telecom standardization (ITU-T);

– radiocommunications (ITU-R);

– telecom development (ITU-D).

The Telecom Standardization Sector contributes, in parallel with other international entities, to defining and standardizing telecommunications systems and to facilitating their interconnection. The recommendations resulting from its activities (ITU-T) refer to multiple national or regional authorities and define specifications for terminals, networks or services.

The Telecom Development Sector manages activities of a more economic and political nature. Its missions are mainly aimed at promoting the development of telecommunications in the developing countries.

The Radiocommunication Sector deals with the management of the radio spectrum. Its activities are in continuation with the conferences of 1906 and all those following. It retains its normative relevance, notwithstanding the current telecommunications industry liberalization. Its legal weight is the strongest among the other sectors with the Radio Regulations. It should be underlined once again that this sector, while within the International Telecommunication Union, extends its competence well beyond telecommunications. Its activity concerns all radioelectric services, as illustrated in the definition of telecommunications services retained by the Regulations and which will be presented below.

The Radiocommunication Sector itself is organized into specialized bodies:

– World and Regional Radiocommunication Conferences (WRC and RRC);

– Radiocommunication Assemblies and their Advisory and Study Groups;

– the Radio Regulation Board (RRB);

– the Radiocommunication Bureau (BR) and its Director.

The essential purposes of the sector are on one side the publication and updating of radiocommunication regulations and on the other side the management of procedures related to the registration of frequency assignments. These two activities have been at the heart of international cooperation since the beginning of radio services. It should be noted that the assignment of frequencies nowadays includes the assignment of orbital resources related to space systems.

The Radiocommunication Assemblies and the Study Groups which they supervise all provide technical elements needed for effective management of the radio spectrum and space orbits, as well as those needed for a harmonized development of radioelectric services. They make recommendations and edit reports that are not legally binding but rather express the "state of the art". These documents serve as reference tools, having been formally approved by the Assembly or at least by the best experts in the field within the study groups. Before it meets, the Assembly submits a report to the World Radiocommunication Conference on the subjects of its agenda.

On this basis, the World or Regional Conference adapts the Radio Regulations and all regulations specific to an area such as the regional plans. The World Radiocommunication Conferences meet approximately every three years on the basis of an agenda established by the ITU Council. They formally approve the adaptations of the Regulations, and those consequently acquire legal value. The signature of the president of every delegation engages his country.

The Radio Regulation Board (RRB) elaborates the rules of procedure necessary to register the frequency assignments whose respect guarantees the Member State rights to use frequencies and orbits. These rules are drafted by the Director of the Radiocommunication Bureau and are approved by the Board. The Director also helps to clear up unforeseen concrete cases which require an interpretation of the Regulations.

Finally the Radiocommunication Bureau (BR), under the authority of its Director, is in charge of recording the assignments and coordinating the implementation of the corresponding procedures. He manages in particular the FRIF, the Frequency Reference International File, or MIFR, Master International Frequency Register, which gathers all the assignments recognized at international level. It plays the role of technical secretary for exchanges of informative or procedural documents:

– maps;

– lists of space stations;

– lists of naval and coastal stations;

– BR International Frequency Information Circular (BR IFIC) which provides information on ongoing assignment and recording of frequencies;

– special sections which describe the requests for coordination in order to determine the countries whose opinions are necessary;

– handbooks and other publications.

The BR and its director are also charged with the general administrative support of the Radiocommunication Sector structures: as secretary for the Assemblies and RRB. It participates in the organization of world and regional conferences.

In addition to the sectors, whose vocation is specialized, several horizontal bodies ensure the overall coordination of the ITU.

The Plenipotentiary Conference is the top policy-making body of the International Telecommunication Union (ITU). It ensures the respect of the Constitution and Convention. The Conferences take place every four years. The Plenipotentiary Conference sets the Union's general policies, adopts a four year strategic and financial plan and elects the senior management team of the organization, the members of the Council and the members of the RRB. The Conference has the competence to deal with any question of general organization, policy and activities. It is assisted by the Council which meets once a year at the Union headquarters in Geneva and ensures the permanent supervision of the organization.

The ITU General Secretariat manages the operational, administrative and financial aspects of the Union activities. The Secretary General directs the General Secretariat.

The General Secretariat's work is to manage the Union. It organizes the regional and world assemblies and conferences, with a number of other meetings. It provides organizational and logistical support with administrative services, documentation, translation and interpretation services in the six working languages of the Union. The General Secretariat informs the actors of the telecommunications industry as well as public opinion by forums or seminars. The Secretary General is also in charge of relations with other international organizations

Established as a specialized United Nations Agency, the International Telecommunication Union is an international organization of Member States. The UN designates them as "Members", with a capital letter, and they enjoy the exclusive right to participate, deliberate and vote. Other entities can take part in activities as "members", such as international organizations or with the approval of one or more Members. ITU Membership in 2006 includes 191 States, over 640 Sector members and over 130 associates. In addition, the national delegations participating in the conferences, assemblies or working groups can nominate experts of very diverse origins to be included as national representatives. The principal bodies of the Union are directed by those elected by the Plenipotentiary Conference, in particular the Secretary General and the Director of the Radiocommunication Bureau. Finally, the institutional committees and the Council of the ITU are composed of qualified people, also reflecting a balance between world regions.

11.2. Radio Regulations

The Radio Regulations are the main reference document, with international significance, for the spectrum management. All national regulations rest on this document and in different situations, in particular with regard to space services, it is the operational instrument directly controlling management actions.

It is a relatively complex and bulky document, only accessible to experts, whose structure and content reflects one century of improvements and successive additions. At every World Conference, modifications are made to this instrument after highly comprehensive technical debates. The outcome is a text with over elaborated tables where any word is weighted, where each formulation reflects a compromise approved by consensus between the Members.

Notwithstanding the precision and range of issues open to discussion, some being accessible only to a handful of specialists, the Regulations are not a type of patchwork. They maintain an overall structure and coherence, which shows a true strategic vision. The negotiations about each particular point in meetings do not exclude bargaining and trade-offs between countries and blocks of countries during the conferences. However, the final document does not keep track of these negotiations. At the end of each conference, it establishes a thorough state of the art which helps to manage the radio spectrum and systems without major hurdles and uncertainty.

The Regulations are published in three languages, French, English and Spanish, the French text being the reference. In their 2004 version, they were made up of four volumes:

– volume 1, whose legal character is the most marked, states the basic provisions which govern the spectrum in the form of "Articles" of Regulation;

– volume 2, known as the volume of "Appendices", gathers some technical appendices which clarify or specify the articles;

– volume 3 presents the "Resolutions" retained by the Conferences which are types of aspirations, decisions or directions for action;

– volume 4, finally, gathers the "Recommendations" resulting from study groups to which the Regulations refer. Through these references, the corresponding recommendations gain a normative character.

Volume 1, which gathers the Articles of the Regulations, deserves a detailed presentation. Here we can find titles of chapters which have represented the "fundamental" principles of management of the spectrum, since the founding.

11.2.1. *The vocabulary of radiocommunications*

The first chapter is dedicated to terminology and general principles. The terminology of radiocommunications is extremely precise and it is often necessary to refer to the word definitions when interpreting provisions of the Regulations. In the inaugural chapter of the Regulations are found lists of definitions which look like those of a contract or a law text which delimit exactly the range of the terms which will be used later. Two examples follow:

– "Telecommunications": any transmission, emission or reception of signs, signals, writings, images and sounds or intelligence of any nature by wire, radio, optical or other electromagnetic systems.

– "Earth station": a station located either on the Earth's surface or within the major portion of the Earth's atmosphere and intended for communication with one or more space stations, or with one or more stations of the same kind by means of one or more reflecting satellites or other objects in space.

Two partial comments will illustrate the operational character of such definitions:

Concerning telecommunications: communications by sound waves are excluded: the "natural" voice or music, of course, but also underwater communications by sonar.

Concerning earth stations: they are all installations on the ground, at sea or in the air and can communicate with an "active" satellite, i.e. nearly all space systems since Sputnik, but are also hypothetical installations which would use a passive relay in space to communicate between them, such as the very first Echo project which implemented a large balloon placed in terrestrial orbit and covered with a reflective film.

The vocabulary thus defined should be preferably re-used in the legal texts of Member States. At least the definitions must remain in conformity. However, administrations commonly use vocabulary enrichments related to specific practices of each State in a spirit of subsidiarity.

11.2.2. *Table of frequency allocations*

The second chapter presents the international table of frequency allocations. This is the central document which organizes the radio spectrum by distributing the resource between the different regulatory services in the three Regions of the world (see Chapter 3, section 3.1). Every band is allocated, as explained in Chapter 7, to primary or secondary services. Adjustments of these tables are among the key stakes of the Conferences. Many allocations are accompanied by footnotes which specify them and mention specific provisions retained by the different countries, without penalizing consequences for other parties. The 2004 version of the table allocates the spectral resource up to 275 GHz.

11.2.3. *Procedures*

The following chapters of the Regulations are devoted to regulatory procedures.

Chapter 3 explains the procedures of coordination, notification and registration of assigned frequencies as well as modification of plans. When respected, they are the basis for the rights of the various countries, in particular the priority rights. Any contentious procedure concerning frequencies must, before anything else, examine whether the procedures were properly carried out.

The description of procedures is very detailed. We find, for example, the list of technical configurations for which international coordination is compulsory, the documents which have to be exchanged with the response times to be satisfied, the role of parties in the instruction of cases and numerous other provisions which imply heavy administrative tasks but in the end are rather natural in their principle. Resulting from these steps is the registration in due form of new assignments in the reference file held by the Radiocommunication Bureau.

Space systems are the object of close attention. They require much forethought and particular precautions. Indeed, taking into account their costs and time of development, satellite systems can be deployed only if their radio configuration is perfectly certain. A satellite operator should not fear technical or legal disputes once the system is manufactured and launched.

Satellite system projects must be made public as soon as possible, at the earliest seven years before the envisaged start and at the latest two years before. This anticipated publication is done in standard forms, on the initiative of the State conducting the project (advising administration). It describes the broad outlines of the system with its major radio characteristics. This publication is announced in a special section of the international information circular on frequencies, the BRIFIC (BR International Frequency Information Circular), which is published every week by the ITU. Subsequently, a dialog between the concerned administrations can begin, with the assistance of the Bureau, and, if necessary, procedures of coordination with systems potentially affected by the new project can be initiated. Any government with the benefit of previously registered assignments can require to be consulted formally by the advising administration and can make any comment, question or objection with the argument of a risk of interfering with a system in place or whose project is already published.

If a satisfactory outcome is duly reached, the project is registered in the international reference file (FRIF) by the BR. This record opens rights to the administration carrying the project and these rights start from the date of the anticipated publication. The advising administration must notify the start-up of the satellites that fall under these rights. If a seven year deadline is exceeded without start-up, the rights lapse.

It should be underlined that in these procedures, the interlocutors of the ITU are the Member States and the acquired rights are a property of the States. It is up to them to determine the legal and contractual mechanisms which will put these rights at the disposal of third parties such as commercial operators for a particular space system, under provisions which are part of the national legislations.

11.2.4. *Interferences*

Whatever attention is paid to define the technical methods for the containment of radio waves, and in spite of the formal procedures applied to implement radio systems, experience shows that prejudicial jamming can occur, i.e. interferences which degrade deeply and more or less permanently the quality of radio channels.

Let us recall that we do not regard an episodically and random jamming, due for example to exceptional propagation conditions as prejudicial interference, likely to create a dispute.

The reasons for prejudicial disturbances can be various, from a purely technical flaw to non-authorized stations. One day, an authorized link that functioned until thus in a satisfactory manner, becomes noisy. It may even become unworkable permanently or during significant lapses of time. The technical analysis of the case shows that jamming is caused by a given transmitter. If the jamming and scrambled stations concern the same administration, it is its responsibility to manage the dispute. On the contrary, if two administrations are involved, they must find together the means to interrupt the trouble. Regulations contribute to help them.

Indeed, the Regulations provisions to deal with such situations with international dimensions, are not technical but they specify the role of implied States, establish their responsibility and propose means to act in concert to resolve the problem. The ITU is not a court, which would judge complaints, but an authority whose purpose is to facilitate the cooperation of States, including leveling litigations between them. The Constitution of the ITU specifies that the ITU coordinates the administration efforts to eliminate prejudicial jamming between the radiocommunication stations of the various countries, and nothing more. It should be recalled that when two radio system operators are concerned by an interference problem, being under the governance of different countries, it is the responsibility of these two administrations to clear up the case with the obligation for each of them to exert its authority on the operators to make them comply with regulations. Indeed, from the ITU's point of view, assignments of frequencies that state the rights of priority are registered in the name of the States: the ITU ignores the operators.

The concerned administrations start an investigation to understand the reasons of disturbance and, on the basis of rules of management, try to agree on a technical and regulatory analysis which determines responsibilities. Each operator must respect the technical specifications applicable to his service. In addition, rights of priority prevail from the dates of assignment and registration, in accordance with the necessary preliminary coordination procedures. If the responsibilities are well established and recognized, the litigation can end as soon as the litigious emission stops.

However, the conclusions of the investigation can be dubious or require a deeper interpretation. Thus, the Radiocommunication Bureau is at the disposal of the Members, to examine the case as a mediator. The parties can refer a cause to the Bureau, with their own analysis and comments. On this basis, the BR formulates conclusions, indicating if possible the entity which in his opinion contravenes the Regulations.

In a nearby area, such as protection against interferences, the ITU organizes the international cooperation for the control of the spectrum through a "system of international control of emissions". Notably, the decametric bands are supervised jointly by the administrations which have monitoring stations for the HF band and want to contribute to this collective task, each carrying out part of an agreed control plan. These stations appear in a Nomenclature of stations of international control of emissions. Each State indicates a central office that will be the interlocutor for its partners for the control of the spectrum and instruction of complaints for interference (see Chapter 9, section 9.1).

11.2.5. *Administrative provisions and provisions relating to services and networks*

The Regulations state various technical and administrative rules to be implemented for the best radio services and networks.

We will not enter into the specialized sections of the regulations which are extremely detailed. As an example concerning administrative provisions, a general rule may be found which obliges the radio channel users to have an authorization (license) to transmit. It may be an individual, personalized or collective authorization to use some common bands. For particular services, a principle of identification is set, making it possible to determine who is emitting. This possibility implies the attribution of standardized codes depending on services.

The regulated technical provisions are even more various, aiming mainly to fix "strict" limits which allow the electromagnetic compatibility of different services in the same band. We can quote the following provisions, some of which have already been mentioned in Chapter 7, section 7.4:

– limits of EIRP applicable to terrestrial stations in a direction close to the geostationary orbit;

– limits to earth stations not to transmit at a low elevation;

– limits of power flux density produced by space stations depending on the wave incidence angle on the ground;

– tolerances on orbital position of space stations and pointing precision of their antennae;

– limitation of earth station EIRP apart from their main beam axis;

– etc.

The Regulations also list provisions applicable to particular services to give them an international statute. This is the case for amateur services.

11.2.6. *Safety: maritime and aeronautical services*

From their beginnings, radio services have played a major role in safeguarding human life. For a long time they provided the only means of communication with ships at open sea and MF and HF connections were very quickly implemented to ensure distress calls. It was mentioned above that, in 1906, the Berlin conference standardized the telegraphic signal "SOS" which has symbolized the call for assistance since then. The shipwreck of the Titanic in 1912 resulted in the harmonization of international frequencies for distress signals (wavelengths 300 m and 600 m). Then, successive conferences regularly improved the provisions of a true permanent service for radio maritime safety, combining the mobile stations of ships and fixed coastal stations. In the same manner, as soon as the demonstration was made of radio connections with airplanes, during World War I, this technique for air safety was considered for use.

The ITU has always remained attentive to these concerns and the Regulations include several chapters dedicated to these subjects.

Maritime and aeronautical radiocommunications are naturally international, in the sense that a ship or plane is sometimes in communication with one country, sometimes with another. The radio rules of use must thus be codified in a uniform way throughout the world so that, whatever the place, communications can be established between people who are not necessarily accustomed to working together. In addition, because of their operational functions, of their use for safety and because of a frequency scarcity, maritime or aeronautical radio exchanges must be fast. They are strictly formatted and respect a universal syntax in order to minimize uncertainties.

The provisions adopted by the ITU for the management of such radiocommunications are part of more general regulations applicable to maritime or aeronautical matters defined by specialized organizations, the IMO (International Maritime Organization) and the ICAO (International Civil Aviation Organization). The dialog between the ITU and these organizations is constant and particular attention is paid to a coherence of work between them. For example, the GMDSS (Global Maritime Distress and Safety System) whose technical provisions appear in the Regulations, has its roots in the SOLAS convention (convention on the safety of lives at sea) whose first version was adopted by the IMO in 1960 and which has been updated several times since.

The provisions which appear in the corresponding chapters of the Regulations, state principles which strongly affirm the professionalism of the safety communications service, at sea and in the air.

First of all, a hierarchy of the communications is described, which involves priorities:

– communications of distress when the total safety of the ship or plane is concerned;

– communications of urgency relating, for example, to serious damage or the health of a person;

– communications of safety announcing information of collective interest for the safety of navigation in a zone;

– other communications.

The Regulations establish the need for an operator certificate, more or less highly qualified, for the use of telecommunications equipment. They require a periodical inspection of the technical equipment in order to be ensured of their conformity. They organize communication monitoring, with mandatory breaks.

Long technical developments are devoted to specify the parameters applicable to various modes of communication according to frequency bands as well as uses. We will not be surprised to learn that communications in Morse code always head the list of services usable for the benefit of maritime safety.

Finally, the procedures for an urgency or safety distress call are detailed in order to organize the exchanges in an effective, always identical, way. The operators must be familiar with the technical rules of radiocommunications but also with the applicable routines of the procedure.

11.3. Assemblies and conferences

The Radiocommunication Regulations are the outcome of long preliminary work prepared chiefly by the assemblies and conferences.

The Radiocommunication Assembly operates study groups, having a mandate to work on topics of interest for radiocommunications, either to make progress to reach the state of the art or to meet the needs of the conference in order to improve Regulations. The study groups cover general subjects such as radio wave propagation but also more specific points. For example, there is a commission dealing with the fixed satellite service, another considering broadcasting and so on.

Study groups examine contributions proposed by governments or other competent entities, regarding issues on the agenda. On the basis of these partial documents, after analysis, discussion and synthesis, they work out a draft recommendation or report which can thus be approved by the Assembly, thus gaining official status. The work plan is codified so that the production of documents functions smoothly and reflects a well-elaborated consensus view at each level.

The Radiocommunication Assembly plans its work in agreement with the World Conferences (WRC). In particular, a meeting of the Assembly, preparatory to each World Conference, known as Conference Preparatory Meeting (CPM), consolidates the recommendations and draws up a report which summaries the various points and proposals about them. It is also this same CPM which, after a WRC, works out the work plan of the Assembly for the following World Conference.

World Radiocommunication Conferences (WRC) take place every three years, on average, and constitute the essential forum for radioelectric spectrum management. The purpose of these conferences is to perform a periodic revision of the Regulations in order to adapt them to the requirements of technical progress. They are a major opportunity to encourage innovation and improve the efficiency of the spectrum, keeping a careful view of the essential rights of actors: system operators and States. Several thousands of participants gather for approximately one month in these general consultation meetings for spectrum matters, in Geneva or another large city of the world. The agenda items are generally prepared at national and especially regional level.

More and more often the topics registered in the program of the WRC are studied within the framework of regional organizations such as CEPT in Europe or CITEL in America which endeavor to work out the views of their members on every point. Thus the debates, at the WRC, often proceed between "geopolitical" blocks which collectively defend positions elaborated and negotiated at their level, rather than between individual States. Furthermore, many preliminary contacts between these regional organizations make it possible to know arguments about the debated issues and often to work out, before a conference, terms of a consensus. This careful preparation does not always prevent confrontations or bargaining, but avoids fanning the debate on a quantity of subjects. The conferences can theoretically resort to a vote to decide on points of controversy, but reaching a consensus remains the rule. Indeed, the logic of the ITU action is based on the voluntary cooperation between States which would be jeopardized by a constraining vote. It is thus necessary for every party to negotiate by showing the force of its arguments and its support, to work out a compromise, to convince and gather the undecided, sometimes to compensate for a profit on an item by the loss on another. In all cases, a perfect control of the forms, structures and mechanisms of dialog ("comitology") prove to be essential.

The development of wireless services has caused, in recent years, inflation in the number of issues entered on the agenda of the successive WRC. Every WRC leaves the following conference a catalog of open subjects and new ones to be discussed. These subjects are examined by the Council for the ITU which sets the agenda for the following conference, at least two years before its meeting. The importance of the points is obviously unequal, from adjustments to improve existing provisions, to strategic decisions, allowing or prohibiting the development of a new service on a worldwide scale.

11.4. Themes of recent interest

All radiocommunication services propose issues which are debated at the WRC. However, for some time, we clearly saw special attention paid by the ITU to space services. This preference is due to some extent to the technological and commercial dynamism of this industry but also to the very particular role of the ITU as "regulator" of this field above the States, since satellite networks are essentially trans-national and use, in any case, orbit-spectrum resources, international goods.

If we consider for example the WRC 2000, held in Istanbul, the major topics discussed were the following:

– identification of bands for the extension of third generation mobile phone networks, known as IMT2000 (mobile service);

– additional spectrum allocation for satellite radio navigation services, for GPS and GALILEO networks mainly;

– conditions to share frequency bands between 11 and 30 GHz, between non-geostationary satellites and other systems (various satellite services, notably fixed satellite service and broadcasting);

– spectrum identification and allocation for high density fixed systems above 30 GHz;

– planning of satellite broadcasting at 12 GHz, for Regions 1 and 3.

On the whole, three subjects out of five were related to satellite systems.

The WRC 2003, held in Geneva, presented the same profile, with the following main topics:

– adjustment of applicable parameters in the bands allocated to navigation by satellite;

– identification of bands close to 5 GHz for the development of radio local area networks;

– conditions for use of bands 13,75-14 GHz by earth stations for fixed satellite services;

– new frequency allocation for downlinks of high density fixed satellite services;

– disturbances brought by satellites with strongly elliptic orbits to various ground services;

– use by ships and planes of systems relevant from the fixed satellite service to offer mobile services.

Again, five subjects out of six concern satellite systems.

The previous world conferences, WRC 1995 and WRC 1999, both held in Geneva, had been strongly marked by the appearance of large projects of non-geostationary satellite networks. They were mobile satellite systems in low orbits (IRIDIUM, thus GLOBALSTAR and others) or systems for the fixed satellite service (TELEDESIC, thus SKYBRIDGE and others), all of them with considerable ambitions and requirements: nothing less than covering the whole planet to offer a direct service to every user, requiring spectrum allocation of several hundred megahertz. However, most projects did not get off the ground.

Generally, considering the relative economical weight of satellite and ground systems, it can be estimated that the former has benefited in recent years from privileged conditions of access to the spectrum (noting also that satellite operators do not pay for their frequencies, at least for the space segment). In any case, satellite systems since their origin have enjoyed close attention from the ITU, constraining the ground systems to cope with an increased sharing of bands. A kind of tacit rule was established which admits that every radio terrestrial system can be supplemented by a homologous satellite system, leading to a certain sharing of the markets: the ground systems provide a density of service at a low cost, the satellite systems provide universal coverage.

In addition to space services, the major trend has been the extension of mobile services which have drawn the ITU-R attention for 20 years. With respect to this revolution of mobile communications, notably the mobile phone, the organization has largely anticipated the needs and opened the way by making available the necessary frequency bands. Until the 1980s, the civil radiotelephone was a luxury, cumbersome product which implemented not very efficient analog standards. Also, the spectrum allocated to the service remained limited, with only a few tens of megahertz, mainly around 150 MHz and 450 MHz. This situation lasted for a long time with the stagnation of the service. However, the cellular networks that appeared around 1980 caused the awakening of the market and, in particular, the Scandinavian system NMT, functioning in the bands 453-457.5 and 463-467.5 MHz, proved to meet mass demand.

Consequently the ITU decided to allocate frequency bands appropriate to the forecast development and deployment of mobile new technologies. The meetings in Geneva (1983), Torremolinos (1992), Istanbul (2000), made successive progress in facilitating the deployment of successive generations of wireless telephony. Gradually, hundreds of megahertz were made available to the mobile terrestrial service in bands around 900 MHz, thus around 2 GHz and 2.5 GHz, with alternatives depending on the Region. It should indeed be stressed that the terrestrial mobile phone services did not find a way for world harmonization and that the main countries or regions did not manage to agree on common bands and standards, particularly as the development strategies differ from one zone to another. This situation did not prevent the mobile from becoming a world success, with a development of multiband and multi-protocol terminals. However, we can complain about these international divergences which complicate the spectrum planning and prevent the true universality of mobiles.

In the future, ideas concerning the flexibility of spectrum management and specifications of self-adapting systems without assignment, as mentioned in Chapter 10, may inspire new proposals such as wireless local area networks which could also provide mobile communications.

11.5. Short-term prospects for action

The World Radiocommunication Conference held in Geneva in 2003 adopted an agenda for the 2007 World Radiocommunication Conference which summarizes the present matters of importance concerning radiocommunications. Among a number of items, the following may be pointed out.

A major issue concerns the new frequency bands which should be allocated to mobile services. Europe wants to secure extension bands for its 3G system (2.5-2.690 GHz) and to identify new bands suitable for more extensive coverage (UHF band) or higher capacity mobile networks (about 4 GHz).

Important debates will take place on interesting aeronautical services. Allocation of new frequency bands to cover the increase of aircraft communication traffic and make possible high bit rate aeronautical telemetry is expected.

The conference will also study the protection and extension of scientific aerospace services which are of fundamental importance for climate and meteorological researches. Earth exploration satellite (passive) services, space research (passive) services and the meteorological satellite services are concerned.

Allocations to all services in the HF bands between 4 MHz and 10 MHz will be reviewed, taking account the new modulation technique impact and spectrum requirements for HF broadcasting.

The conference took place in Geneva in the fall of 2007.

Chapter 12

Regional Bodies

Intermediate entities, such as regional harmonization institutions and supra-national political structures, have appeared recently and placed themselves between the International Telecommunication Union and the States. Their activity on spectrum management has been steadily growing. The economic "globalization" and the general development of people and business mobility among the national actors has naturally led to a strengthening of their cooperative role in order to trigger concerted actions on geopolitical zones of the planet.

Among the regional institutions which play an important role, can be mentioned the European Conference of Postal and Telecommunications Administrations (CEPT), the Comision Interamericana de Telecommunicaciones (CITEL), and the Asia-Pacific Telecommunity (APT). Other organizations have also emerged to deal with issues regarding the Arab and African countries.

Among the supra-national political structures showing notable ambition for frequency management, we can quote for example the European Union (EU). The implication of the EU in spectrum management is an important innovation since it is not a Member of the ITU but it does have true prescriptive power concerning frequencies with respect to its Member States.

To illustrate such regional harmonization activities throughout the world, let us introduce the European structures, CEPT and EU, as well as the CITEL and APT.

12.1. The European Conference of Postal and Telecommunications Administrations (CEPT)

The European Conference of Postal and Telecommunications Administrations (CEPT) was created in Montreux, Switzerland, in 1959. It was established between the administrations of the postal and telecommunications services of nineteen European countries for cooperation in the fields of competence of its members. Before the creation of CEPT, some countries envisaged the ambition of a true European institution that would manage a common European domain of postal and telecommunications services, but they quickly renounced this too ambitious and premature objective and preferred to create an assembly for dialog, with a stable status and formal structures. A "conference", open to the postal and telecommunications administrations of European states with a very light organization, was considered to be more realistic than a political structure. Consequently, the future work of this conference could succeed only with recommendations having no legal or constraining character, reflecting however consensus between professionals and consequently suitable to be transformed into projects. Since the beginning of this institution, it was also desired to develop, within CEPT, common positions for ITU conferences, and to look further into the international decisions together.

The management of CEPT was entrusted to a "managing administration" on a voluntary basis, chosen among the members of the Conference. A modest coordination office, in charge of tasks of secretariat, communication and filing, was installed in Berne. For a long time, only countries of Western Europe took part in the Conference, the Eastern European countries, thus under Soviet influence, did not wish to participate.

Radio spectrum management had been mentioned since 1955 as being likely to interest European countries in order to build a cooperation to organize the use of frequencies. However, at that time, the international HF connections were the main international radio service and it was thought that the importance of radio matters was really marginal with respect to the postal questions, for example.

It is remarkable to look at the evolutions of the past 50 years. Today the former "Eastern European countries" are active members of the Conference which gathers about 47 countries (2007). Wireless services have become the major domain for work, so much so that CEPT seems to be one of the major actors of radiocommunication policy in the world. The member countries are as diverse as France and Russia, Iceland and Malta, the Vatican and Azerbaijan.

During all these years, CEPT has successively gone through various priorities but always in view of telecommunications harmonization in Europe. It has also played a paramount role in the birth of great projects such as the EUTELSAT satellite network and the GSM. It has contributed to world standardization, in particular in the area of digital techniques of telecommunications. In short, it has made it possible for Europe to become a major geopolitical center for telecommunications. However, the standardization activity (main activity in the 1970s) has moved gradually to ETSI (European institute for the standardization of telecommunications) created in 1988, while spectrum regulation and management has continued to be the responsibility of administrations.

CEPT has now set up a permanent technical office, installed in Copenhagen. A European Radiocommunication Office (ERO) was created in 1991 and, since 1994, a European Telecommunications Office (ETO) was added. These two structures merged and formed in 2002 the new ERO. At the same time, CEPT was reshaped by a plenary assembly in September 2001, notably with the creation of a committee for electronic communications, the Electronic Communications Committee (ECC) bringing together former specialized committees for radio and telecommunications.

These changes are not only anecdotes: they illustrate the huge evolution which the ITU has also gone through. Radio becomes the key domain for regulation under control of States and political entities like the European Union. The technical standardization and management of licenses is now a concern for operators on the market and independent regulators. The members of CEPT, being States, focus their activity on the governmental matters, mainly the management of the spectrum. These modifications logically go along with the new regulatory framework that the European Union is aiming towards, to foster the evolution of the information society, the development of electronic communications and the convergence of digital technologies.

Concerning radio, the activity of CEPT is performed via special working groups in the ECC:

– the FM group deals with spectrum planning, with a constant view on harmonizing the use of frequencies in Europe;

– the SE group undertakes all studies of electromagnetic compatibility which spectrum planning requires. It proposes technical limits which will have to be imposed on radio emissions for a good compatibility;

– the RA group examines all legal or economic questions concerning radio.

A particular group, the CPG (Conference preparatory group), prepares contributions by CEPT at international conferences. It develops a catalog of common European proposals (ECP) which will be presented collectively at the ITU as contributions to the agenda of the coming World Radiocommunication

Conference (WRC). The long internal debate in advance of the conference, held on each subject within CEPT, makes it possible to draft a first synthesis between very diverse European countries and positions. It anticipates, to a large extent, the debates which will take place at the ITU, which gives a particular relevance to the positions retained by the CPG. This is still truer as CEPT attempts to take into account the views of other regional bodies and to maintain close links with other geopolitical zones, especially those well known to CEPT, such as Africa and the Middle East. In this context, the capacity of rallying other countries to the CEPT thesis is confirmed at every world conference.

CEPT does not have legal capacities and thus cannot establish regulations or documents with a legal binding character: it has already been mentioned that it was a forum for voluntary cooperation between administrations. However, the conference drafts formal decisions, subject to public evaluation and comments. The administrations are thus invited to approve these decisions and commit themselves to implementing them within pre-determined deadlines. Such a mechanism, which respects the freedom of the States, is highly incentive since it highlights the administrations which refuse to go along with common outlooks and constrains them to strongly justify their objections. In fact, the authority gained by CEPT is so high in Europe that practically any decision by the States, in regard to spectrum, is made with reference to its work and is almost always in conformity with its directions.

The permanent office of CEPT (ERO) studies some prospective issues and presents reports on strategic topics. It keeps the documentation of CEPT up to date, notably decisions but also a European table of frequencies (EFIS) which brings together the national tables and proposes the outline of a European harmonized table to aim at in the future. During recent years, CEPT was also requested, by mutual agreement, to manage practical actions making expensive technical instruments available to the community of its members. For example, particular conventions (MoU, Memorandum of understanding) were signed between various members to produce and manage, via ERO, a simulator of radioelectric environment (SEAMCAT) or to jointly use the satellite monitoring station at Leeheim, in Germany. The office's activity and management are supervised by a council.

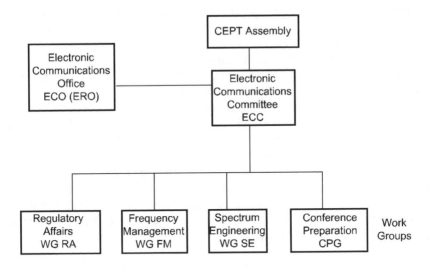

Figure 12.1. *CEPT organization for spectrum management*

12.2. European Union

In its efforts to develop a European common internal market and to promote service harmonization in its territory, the European Union, particularly the European Commission, has encountered radio resource management. Indeed it recognized that controlling this resource predetermines the deployment of a number of networks and modern systems of electronic communication. In this respect, the extraordinary success of mobile GSM was a revelation for many.

Thus, the EU has sought for a few years to build a competence in this field. In particular, the Directorate-General of the Information Society of the European Commission has impelled a number of outstanding actions concerning radiocommunication networks. It recognized the benefits of a constructive partnership with CEPT and took part as an observer in its work. It decided to build structures adapted to an active dialog with the main actors of the market and the national regulators.

For some time now, the EU directions have been formalized in reference texts and thus frame the regulation and the practice of the Member State administrations. A directive of June 25, 1987, concerning the deployment of GSM networks and a decision of December 14, 1998, on third generation mobile phone networks, showed, with force, the will of the EU to put in place harmonized networks of radiocommunications.

In this respect, the voluntary approach to favor a technical standard, UMTS, to structure the networks of third generation mobile telephony and to standardize the corresponding frequency bands on the European territory by enforcing the recommendations of CEPT, was a major political act of vision. Incidentally, it provoked the hostility of the American administration which deemed this a distortion of free competition rules. Such a prescriptive policy, directly inspired from the GSM success, must however prove, each time to be effective on the new market concerned. It is in fact not obvious that all the radio networks and services can be built on the same regulatory basis. There are certainly applications and services which deserve an *a priori* alignment, as GSM proved, while others feel better in free confrontation on an open market. The management of spectrum must learn on a case-by-case basis to distinguish between the winning strategies and the non-productive ones. Indeed, the experiment shows that several directives of strategic planning were relative failures, for example the directive of 1990 relating to the ERMES harmonization of paging.

The technical decisions are not the only ones to influence the market dynamics. In the case of UMTS for example, whose deployment was to be made in the States of the European Union about January 1st 2002, it was obviously too early, as the rules for license attribution were not yet fixed. From this situation, a true cacophony resulted between the States, some recommending the attribution of licenses by comparative tender of offers in conformity with a set of conditions (beauty contest), others retaining auction procedures. This led in some cases to licenses with a very high value and to others with a trifling value. These unmatched and not synchronized steps, around a project whose market was still dubious, contributed to the destabilization of several important operators, in particular their financial balance, without any benefit for either the economic activity or the consumers.

We can stress that uncertainties on the good governance of the spectrum, as expressed by the case of UMTS, are closely related to a certain image of frequencies, namely as a scarce resource which should be shared with fixed quotas, assigned in a more or less exclusive way. These views thus call for a rigorous formalism in the license attribution, in order not to raise criticism. The Green Paper by the European Commission, which we will discuss later, talks about an "essential but increasingly scarce resource". Such a rigidity and solemnity of top-down procedures may sometimes be contradictory to the market reality. In reaction to this approach, a school of thought calls for a standard market-approach to the spectral resource, which would become, within the framework of a primary and secondary market, an object of trade between economic actors, without reference to choices of services or standards. The promoters of this policy hope that such a vision would have the virtue of increasing efficiency and remedying the relative shortage. This issue will be addressed in Chapter 15.

All these debates and other considerations, such as the incoming convergence of telecommunications and audiovisual technologies, have fed reference texts relating to a European strategy for spectrum management. Thus the European Commission published in 1998 a "Green Paper on the policy as regards radio spectrum in the context of the Community policies of telecommunications, broadcasting, transport, research and development". This document summarized the policy pursued until then by the Commission and submitted to public discussion some directions for the future. Following the discussions, the EU adopted in March 2002 a "Decision relating to a legal framework for policy as regards radio spectrum in the European Community" (767/2002/CE).

Among the objectives of this decision appears the coordination of Member State policies for spectrum availability and effective use. This coordination must necessarily involve harmonization procedures, support for common European positions at the ITU and better mutual information. It justifies the creation of specialized structures for dialog, notably the "Radio Spectrum Committee" which assists the Commission. The decision also recognizes the privileged role of CEPT as a forum for technical negotiations concerning spectrum management, so much that the Commission entrusts it with mandates to work out the substance of directives which may be imposed thereafter on the Member States. This cooperation between those two authorities, one technical, the other political, should prove to be very efficient.

This European situation is interesting as it brings together three types of different actors who cooperate with discipline and flexibility: States, the EU and CEPT (not forgetting ETSI, as a standard provider). A subtle mix of regulation, voluntary commitment and liberty of action makes it possible to work at the same time towards a common radio space and to spare national needs. Other mechanisms such as bilateral cooperation, for example for the frequency coordination at borders, still complete the set of instruments for spectrum management.

It results from such flexibility that the spectrum, in Europe, is organized in harmonized and non-harmonized bands between States. Within the European Union and according to the field of competence of the European Commission, the harmonization policy concerns primarily commercial radiocommunications, mainly for telecommunications and audiovisual sectors. A distinction is clearly made between harmonized and non-harmonized domains. Freedom of movement and the free use of terminals are legal in harmonized bands. In the non-harmonized bands, freedom of movement is also the rule, except particular safeguard measures, but the effective use is regulated by the States according to their frequency allocation tables and the authorizations for frequency utilization which they deliver. Thus, labeling and operating manuals of terminals play an important role in informing consumers. A European directive of 1999, known as the "R&TTE Directive" (radio and telecommunications terminal equipment) describes all these methods concerning "the radio equipment and the final equipment of telecommunications and the mutual recognition of their conformity". This directive mentions some essential requirements that any terminal must respect, notably concerning the health and safety of users.

The Commission pursues a diligent policy to come up with common views on spectrum policy among member states. Among recent documents accessible on the European Commission website and dealing with radio spectrum policy, let us mention:

http://europa.eu.int/information_society/policy/radio_spectrum/index_en.htm

Communication "A forward-looking spectrum policy for the European Union – second annual report", COM (2005) 411

http://europa.eu.int/eur-
lex/lex/LexUriServ/LexUriServ.do?uri=CELEX:52005DC0411:EN:NOT

Communication "A market-based approach to spectrum management in the European Union", COM(2005) 400

12.3. Other regional structures

Other regional structures make administrations cooperate throughout the world although they have apparently played until now a more minor role than CEPT in Europe.

CITEL

The inter-American commission of telecommunications (Comision Interamericana de Telecommunicaciones or CITEL), instituted with its present structure in 1993, depends on the Organization of the American States whose seat is in Washington DC in the USA. It gathers the representatives of 34 Member States and about 200 associate members who take part in the debates but without voting rights. Its role is to promote the cooperation of the Member States in the field of telecommunications, to harmonize the networks and services, to facilitate communications on the American continent and to prepare collective positions for the ITU conferences. The basic work is done by permanent Consultative Committees, of which one is for radiocommunications. They report to a permanent executive committee of 11 members. A secretariat performs daily operations.

The permanent Consultative Committee for Radiocommunications is competent on all subjects concerning the planning of the spectrum and the orbit-spectrum resource for satellite systems, particularly in relation to working groups and conferences of the ITU. It can address any subject concerning the operational management of the radio resource which interests the community of the Member States.

The goals pursued by the committee are rather straightforward:

– to contribute to a harmonized use of frequencies and services in the countries of the American continent, to guarantee the best overall electromagnetic compatibility therefor;

– to support the deployment of new technologies which improve the spectral effectiveness;

– to prepare in cooperation the main conferences and meetings of the ITU, in particular by working out common proposals (IAP).

According to these perspectives, specialized working groups pursue the study of various subjects:

– evolution of bands devoted to various regulatory services;

– preparation of ITU conferences;

– procedures for spectrum use reorganization, etc.

Other permanent committees can also make provisions concerning the spectrum, such as the committee of broadcasting experts.

APT

The Asia-Pacific Telecommunity (APT) aims to inspire a true policy of regional development by promoting information technologies in its area. Its influence is on the rise, commensurate with the growing economy of the region.

Created in 1979 by an intergovernmental agreement, the APT has 32 Members, four associated members and 100 partner members. The objective of the Telecommunity is to facilitate the development of telecommunications services and electronic communication infrastructures among its members, in particular in the less advanced zones.

The topics open to dialog within the APT are very wide, going well beyond concerns of technical harmonization, such as the promotion of the information society, technology transfers and development of human resources. However, the ambition to develop a regional cooperation in the fields of radio and standardization and also to reinforce the international weight of the area figures is at the top of the action plan.

A general assembly and a board of management direct and supervise the operation of the APT. The main activities are made up of "programs" according to a very flexible structure, adaptable according to the priorities:

– preparation of world conferences;

– development of third generation mobile phone;

– standardization;

– education and human resources;

– etc.

Collective preparation for the World Radiocommunication Conferences was first carried out in 1996, for the WRC 97, within a group of experts. The definition of joint positions being proven effective, it was decided to renew this policy for the following conferences within a preparation group for the conference (APG).

These regional structures, and others, cooperate in an informal way in the preparation of the WRC by the presence of invited mutual observers. In addition many organizations with broader vocations than radiocommunications take part in the debates of their competences and interests.

Chapter 13

National Spectrum Regulators
and Institutional Debates

According to the ITU Radio Regulations framework, the ultimate managers of the spectrum are the States. In line with this task, they publish laws and regulations that establish the principles applicable to electronic communications, of which radioelectric systems are an important part. They also set up the structures and institutions needed to implement these regulations.

Many countries choose distinct structures for the radio spectrum part devoted to governmental activities and for the spectrum part used for commercial or private activities. The first part remains governed by the government or a governmental entity; the other can be governed by an independent authority. Alternatives exist, for example the telecommunications and audio-visual regulators may be distinct or, on the contrary, "convergent". In the same way, according to whether a State is centralized or has a federal structure, the shared responsibilities for frequency management can vary.

Some countries finance the regulators by a subsidy from the State general budget, considering that the fees associated with the use of the spectrum, as public property, must return to the general budget. Others consider that all or part of these fees can finance the regulators, as a remuneration for their activity.

Thus, the design of regulatory institutions at a national level may be very different: each country making its own choices concerning its regulation and the structures which have the responsibility to implement it. This diversity has increased since the 1980s, as countries widely varied in their ambition and schedule for the liberalization of broadcasting services and telecommunications. Many countries, but not all, have put in place independent authorities, or more or less autonomous agencies, created to discharge the government from directly acting in relation to the market. Different sets of management structures, with adapted responsibilities, have resulted, some with the objective to promote a large and fair competition to access and use the spectrum, others maintaining an extensive administrative control over a major part of the resource. As a general principle the national legislator states the basic rules in the form of laws and decrees. The regulator, or regulatory system if different entities are involved, performs the daily operational control and takes decisions needed by economic actors to access and use the spectrum resource.

Until now the legal provisions related to the spectrum management were not generally self-governing. They often appear within different and more general texts that deal with matters of collective interest such as telecommunications or electronic communications, audio-visual activities, space, public safety or health, etc. In these texts the radio spectrum is seen as an instrument for the benefit of particular sector policies and does not have to be regulated for itself, except by the Radio Regulations. This sectored view, however, is being challenged by a growing awareness that the spectrum is a common resource which should be comprehensively understood and managed. This new approach gives rise to transverse concepts and concerns, such as the care for public health with respect to radio waves, the economical value of frequencies and the necessary emphasis on an efficient use of the global resource.

Let us review the contrasted institutional choices of a few countries.

13.1. The USA

The organization of the USA for spectrum regulation reflects the existence of two categories of frequency usages: public and commercial. These two types of uses are managed by different regulators.

"The Communication Act of 1934 established the legal basis for spectrum management in the USA. The act created the Federal Communication Commission (FCC) as the agency responsible for licensing all private sector and non-federal-government use of the radio spectrum."[1] The FCC is one of the oldest independent regulation organizations. It is an independent regulatory agency (that is not part of the executive branch) and its five members are appointed for five-year terms by the President with the advice and consent of the Senate. Although the FCC shares certain regulatory functions with agencies of the individual states, it has exclusive

---

1 For a full description of the role and organization of the FCC: www.fcc.gov.

jurisdiction over non-federal-government spectrum management issues. The FCC carries out its responsibilities through procedures set forth in the act and in more general statutes governing the administrative procedures used by federal agencies. These procedures, generically referred to as rulemaking proceedings, require the agency to notify the public of proposed actions, to allow opportunities for public comment, to provide reasoned, written decisions based upon the public record, and to permit appeals of those decisions to the federal court system.

The act authorizes the FCC to regulate non-federal-government use of the radio spectrum in the public interest, but it reserves for the President the authority to allocate radio frequency bands for use by the federal government itself. The President, in turn, has delegated this responsibility to the secretary of commerce and, thence, to the administrator of the National Telecommunications and Information Administration (NTIA), an agency within the US Department of Commerce, which is an executive branch agency.

The NTIA coordinates the federal government's use of its portion of the radio spectrum with the advice of the Interdepartmental Radio Advisory Committee (IRAC). The procedures used by the NTIA and IRAC, which are deeply technically directed, involve formal coordination with concerned agencies. It is beyond the scope of this book to discuss the spectrum management procedures used by the NTIA and IRAC but it may focus on the procedures used by the FCC in managing non-federal-government use of the spectrum.

The concern for a rigorous management of the spectrum appeared very early in the USA, notably after the shipwreck of the Titanic in 1912. This same year a law established that the use of radio frequencies required an authorization from the federal government. Thus, in 1927, a federal committee for radio (Federal radio operator commission) was established with the mission to manage all the frequencies devoted to microwave links between the States of the Union and international connections. Seven years later, the "Communication Act" created the FCC to which the prerogatives from the federal committee for radio were transferred.

The tasks which the FCC carries out relate to the whole chain of management of the spectrum and frequencies: spectrum planning, technical standardization when appropriate, definition of rules and procedures insuring electromagnetic compatibility between systems and services, licenses, frequency assignment, protection of users against interferences, public safety concerns, etc.

The FCC has played a major role in introducing economic incentives to spectrum efficiency. It very soon expressed concerns relative to flaws and limitations of administered comparative tenders ("beauty contests"), as well as its dissatisfaction with the lotteries which it implemented initially. As soon as 1994, it tested auction methods better adapted, from its point of view, to the commercial approach of radio services. Since then, these procedures have been enlarged and adopted in a number of countries.

The NTIA, created in 1978, is an agency under the US Department of Commerce. It is in charge of the American public interest in debates relevant to the spectrum which can have an impact on defense and public security. It has the responsibility of negotiating with foreign governments, notably within the ITU, the provision of the radio resources necessary for the country and the government of the USA. Like the FCC, the NTIA is concerned with the best use of the spectrum and contributes to direct its evolution. However, this organization is not strictly speaking a regulator, even if it takes into account the spectrum requirements that can profit the commercial world, but rather the guarantor of American political interests and the radio resources available for State purposes.

In the period 1990-2000, the NTIA and the FCC undertook a careful investigation of the methods of spectrum management in order to increase the spectral efficiency and make the resource accessible to the greatest possible number of parties. It seems indeed that a relative dissatisfaction had appeared on this subject in the USA and that the current methods were criticized by the industry. In this respect, in the USA like elsewhere, an organization which divides the spectrum between two agencies is exposed to criticism, in the name of flexibility. We can indeed think that such a separation implies artificially created conflicts which may be reflected in the views expressed in the international organizations by the American administration.

13.2. The UK

The UK pioneered the de-monopolization of the telecommunications sector in Europe. It also set the trend for the establishment of specialized regulation bodies covering mainly the different commercial communication activities. A particular technical institution, the Radiocommunication Agency (RA), managed the spectrum until 2004. These various structures merged into a single entity, Ofcom (Office of Communications) in 2004, providing a test case on whether the convergence of regulation institutions, preached by many, is as efficient as can be expected[2].

An analysis of the missions of the various structures in place before they merge makes it possible to understand the British choice, focused on a better service to the consumers. In addition to technical management concerns, the main objective of the

2 For a full description of the role and organization of Ofcom: www.Ofcom.org.uk.

regulation is indeed the development of the best possible offer to the users and customers of electronic communications: diversity, quality and better cost are top priorities. All tools should be combined for this objective.

In the UK, until 2004, five regulators shared the task in the area of commercial services for telecommunications and broadcasting; the governmental communications, notably for Defense, were managed, as usual, by the various competent ministries.

Three regulators worked on audio-visual issues in accordance with the law on audio-visual services of 1996:

– the "Broadcasting Standards Commission" had to define the general standards applicable to audio-visual services: technical standards for coding and modulation, quality standards, implementation rules. This mainly technical organization guaranteed the coherence of the service offered on various infrastructures: radio, cable, satellite;

– the "Independent Television Committee" (ITC) attributed the authorizations of emission (licenses) to commercial radio televisions other than public televisions. It controlled their programs, took care of their engagements and maintained equity of their development conditions. As an authority independent of the government, the ITC was financed by the amount of the licenses. It planned the frequency bands for television;

– the "Radio Operator Authority" was the equivalent of the ITC for independent radio.

For telecommunications, a dedicated regulator, the "Office of Telecommunications" (OFTEL), was created in 1984, corresponding to the end of the British Telecom monopoly. Many new operators entered the various market segments thereafter. OFTEL, an independent authority, was intended to insure the quality of telecommunications in the country, i.e. to guarantee the best service availability, a diversity of choices offered to the customers, at the right price. The main instruments to attain these objectives were legal and economical, with the office examining user complaints and encouraging competition between operators. OFTEL had to manage many cases concerning radiocommunication networks but without having the control of their spectrum resource which was a Radiocommunication Agency responsibility.

As already mentioned, the radio spectrum regulator in the UK, for the non-governmental applications, was the Radiocommunication Agency. This agency reported to the Ministry of Trade and Industry. Spectrum valuation was one of the RA's favorite topics. This institution strongly contributed to the economic works on the spectrum value in relation to its relative scarcity and the competing advantage which a user gets from frequency band attribution. In this respect the implementation of auctions for the attribution of the third generation radiotelephone bands (UMTS) was a first in Europe, at the beginning of 2000. Other works such as those concerning the resale of the spectrum (secondary market) also contributed to a revelation on the economic value of this immaterial goods.

From January 1, 2004, these five institutions for the regulation of electronic communications in the UK merged into the Office of Communications (Ofcom) which inherited all their former responsibilities and missions. It was initially established under the Office of Communications Act 2002 but received its formal accreditation under the Communications Act 2003[3].

In carrying out those duties, Ofcom is to:

– ensure effective delivery of services to customers that meet their needs;

– offer an integrated approach to communication regulation;

– provide consistency of regulatory approach;

– deliver clarity of policy;

– have the ability to respond flexibly;

– strive to maintain diversity and quality within the communication sector;

– promote the expertise of existing regulators;

– attempt to deliver faster resolution of complaints;

– issue licenses and make policy decisions.

A major part of Ofcom's duties concerns the radio spectrum which it licenses for a range of activities including TV, radio, mobile telephony and private communications. Some licenses are granted subject to an administrative fee while others are allocated via auctions. Procedures have not been much affected by the transition from former regulators to Ofcom.

Ofcom may look like FCC. Where it differs from the FCC's historic approach is that Ofcom is technology neutral and instead regulates services. As an example, concerning broadband services, the FCC regulatory approach required a distinction to be made between "telecom services" and "information services", whereas Ofcom makes no distinction between broadband via DSL, cable modem or satellite.

The idea is that Ofcom should favor light-handed regulation but intervene where necessary and, in so doing, create conditions in which dynamic industries can

---

3 The following sections are based on an unpublished note by Peter Curwen.

develop. Insofar that the public interest is to be protected, competition rather than regulation is seen to be the preferred response.

Where a company has no significant market power, Ofcom's ability to make sector specific rules is limited to essential issues such as consumer protection, access and interconnection rights, special planning requirements, rights to use telephone numbers and spectrum, access to emergency services and administrative charges.

Where companies have a significant market power, Ofcom may deal with non-discriminatory interconnection and cost-based pricing for interconnection, requirements for vertically integrated companies to produce separate regulatory accounts, rules against unfair cross-subsidies and rules prohibiting undue discrimination or undue preference between the company's own business and that of third parties.

When creating Ofcom, benefits were expected to flow from the greater scale of the new regulator. Benefits were also expected to arise from a new service and market-oriented culture, the introduction of best practice and the simplification of legacy approaches. On an ongoing basis, Ofcom was to be financed out of spectrum license fees, broadcast license fees and income from authorizations for networks and services.

13.3. France

In France, for a very long time and as in most countries, administration had a near monopoly of the spectrum use. This control was considered natural as it remains today for national sovereignty services such as defense or public security, since telecommunications and audio-visual broadcasting were State monopolies.

The situation changed during the 1980s with the opening of radio services to competition and the entry of private operators. For broadcasting, successive laws in 1982 and 1986 authorized private local radios and created an independent regulator, the "Haute Autorité", now the "Conseil Supérieur de l'Audiovisuel" (CSA). As a strong indication of change, the law of 1982 proclaimed, in its article 1: "audio-visual communication is free". Television privatization followed. A new national landscape for audio-visual sector had been created.

The law of 1986 gave a clear statute to the radio spectrum by stating (article 22) that "the use by the holders of authorizations of radio frequencies available on the territory of the Republic constitutes a privative mode of occupation of the public domain of the State". The law also instituted the concept of wholesale spectrum public managers (now called "affectataires") in its article 21: "The Prime Minister defines... the frequency bands or the frequencies which are attributed to the administrations of the State and those whose attribution or assignment is entrusted to the CSA".

Telecommunications were also opened to competition and radio operators appeared, notably for mobile telephones. An independent authority was created and appointed in 1996, the "Autorité de Régulation des Télécommunications" (ART, now ARCEP) to regulate this sector. Another part of the spectrum was attributed to this authority for telecommunications needs. Thus, ARCEP is also an "affectataire", like the CSA.

Concerning spectrum management, the French model is quite original. It is characterized by a technical agency performing the overall planning and management of the spectrum, the Agence Nationale des Fréquences (ANFR), placed under the authority of the Ministry of Industry and the Prime Minister. It draws up and publishes a National Frequency Allocation Table where the frequency bands allocated to different services are determined with their corresponding "affectataires", ARCEP for telecom, CSA for radio and television, and the various competent Ministries (Defense, Interior, etc.) according to their responsibilities. There is, thus, a two-tier institutional setting: band attribution to "affectataires" which, at a second level, assign frequencies to final users in their attributed bands[4].

ANFR

The ANFR objective is to put in place a national, global, consistent, comprehensive, and transparent spectrum management framework. This choice of ANFR which may look like a "national ITU" for "affectataires" is considered a factor of efficiency and flexibility since the spectrum is considered a common resource which can be fluently shared and permanently redistributed between services and "affectataires" with a full transparency. Common management procedures are also provided, such as frequency assignment registrations or coordinations which facilitate the electromagnetic compatibility of systems. It has allowed, for instance, an active policy of re-allocation (re-farming). Frequencies are consequently, shared among nine user entities, the "affectataires", seven ministries (the main user of allocated spectrum being Defense) and the two independent authorities competent in telecommunications and broadcasting. The ANFR considers these "affectataires" as being on the same foot as each other, being of course attentive to the specificities which the law recognizes for each one of them, for example the confidentiality prerogatives from which Defense and public security services benefit.

---

4 For a full institutional description of the role and organization of spectrum management in France: www.anfr.fr.

The separation of telecommunications from broadcasting, with two different "affectataires", was based on the consideration that their reference frameworks for regulation were different. For telecommunications, network accesses, operator and service competition and cost-orientation are the imperatives. For a content industry like radio and TV broadcasting, cultural quality and diversity, political pluralism, ethical standards and protection of children are the main requirements. This has justified distinct, non-convergent institutional bodies.

However, this choice has drawbacks, as it puts aside the spectrum allocated to the audio-visual services and maintains an artificial border between converging services. Recent evolutions, such as TV on mobiles, attenuate this specificity. It seems indeed desirable that the audio-visual sector should be fully integrated into the community of radiocommunication services, at least for technical matters such as spectrum management.

ARCEP (Autorité de Régulation des Communications Electroniques et des Postes)

The Autorité de régulation des télécommunications (ART) was created in 1996, as an independent regulator, in accordance with the process to open telecommunications to competition. It thus became ARCEP (authority for the regulation of electronic communications and post) in 2005, with the inclusion of postal service regulation in its competences. The President of the Republic and the Presidents of the National Assembly and the Senate, nominate the 7-member committee that governs the Authority, each one nominated for six years, which is non-renewable. These provisions guarantee the institution's independence. However, tight institutional relations are established between the regulator and parliamentary committees and some significant decisions need to be approved by the Government. As a result of the pressure from market actors and according to the European Union regulatory framework, the law has recently provided the regulator with greater authority. In France, ARCEP is one of the main radio spectrum actors, after Defence, since it manages all the frequency bands dedicated to radio services for private and commercial telecommunications.

The responsibilities of the Conseil Supérieur de l'Audiovisuel (CSA, High Council for the audiovisual sector), for the management of frequencies of broadcasting services, are rather close to those of ARCEP. However, as mentioned above, the motivations and thus the principles of management are rather different, being mainly inspired by a policy on program contents.

It appears, at least theoretically, that the objectives assigned to the French regulation go beyond pure technical and economic efficiency in spectrum usage. Regulation and spectrum management have to take into account general goals and ambitions such as employment, regional planning or economic development, and social concerns such as a universal service. As an interesting consequence, band attribution procedures based on purely financial criteria (auctions) have not been used. Multi-criteria decision procedures have been preferred.

13.4. Germany

The regulation authority "Bundesnetzagentur" (formerly Regulierungbehörde für Post und Telekommunikation) is responsible for telecommunications, postal services and energy markets. From 2006 it has also been in charge of regulating railway infrastructures. This agency is a Federal Office depending on the Ministry of Economics, being however broadly independent. The composition of the advisory committee, a quite influential institution within the authority, reflects the balance of power between the Parliament and the federal structures of the country.

13.5. Italy

Three bodies are involved in spectrum planning and management in Italy:

– the Ministry of Communications is responsible for spectrum allocation to private and public services, frequency assignment for civil use (except broadcasting services) and representation in relevant international bodies;

– the Ministry of Defense manages the frequencies for military and public security applications;

– an Authority of Telecommunications (AGCom, Autorità per le garanzie nelle comunicazioni) manages the procedures for granting general authorizations and licenses for telecommunications and frequency planning for broadcasting services. AGCom enforces the existing regulation and can implement new provisions to cope with the changing technological environment.

The president of AGCom is appointed by the President of the Republic on the proposal of the Prime Minister approved by the Ministry of Communications. The president as well as the members of commissions and councils are appointed for seven years, non-renewable.

13.6. Asia-Pacific

The most traditional and historical way of managing the spectrum is to have it handled by the ministries in charge of communication matters. This is still the case for big countries in Asia such as China, India, Japan or Indonesia. Branches of different ministries, which keep the spectrum under close governmental control, perform spectrum management.

Thailand, on the contrary, has put in place institutions very similar to some European ones. Its National Telecommunications Commission is in charge of regulating telecommunications networks, including wireless services. A distinct entity is envisaged for broadcasting matters. Australia and New Zealand have featured innovative institutions and policies.

Australia[5]

In 2005, the Australian Broadcasting Authority (ABA) and Australian Communications Authority (ACA) were joined into a single entity in charge of managing the radio spectrum in Australia. Before this, the spectrum for broadcasting services was managed by the ABA under the Broadcasting Service Act 1992, the other spectrum resources being managed by the ACA.

Today there are four main players involved in civil spectrum management and policy developments. These are:

– the Australian Communication and Media Authority (ACMA);

– the Department of Communications, Information Technology and the Arts (DCITA);

– the Australian Competition and Consumer Commission (ACCC);

– the Minister of Communications, IT and the Arts.

The ACMA is the agency in charge of spectrum management. It also has a policy advisory role for the government which it shares with the DCITA. Many activities require both party contributions before they can proceed. The Minister's competence includes the designation of spectrum bands for broadcasting purposes, the determination of competition limits to apply for spectrum assignment to operators (i.e. the spectrum amount that specific bidders can obtain), the spectrum re-allocation decisions and renewal of licenses (this can also be done by ACMA).

The ACCC is the authority responsible for the regulation of access to essential facilities of national importance, principally the utility network industries. Australia has introduced market mechanisms for spectrum assignment (auctions and trading).

---

5 This section is based on Country Studies by Mark Scanlan, in: J. Scott Marcus, Lorenz Nett, Mark Scanlan, Ulrich Stumpf, Martin Cave, Gerard Pogorel, *Towards More Flexible Spectrum Regulation*, Wik-Consult, Study for the Federal Network Agency (Germany), December 2005.

The ACCC's merger rules contained in the Trade Practices Act 1974 (TPA) are applicable to primary and secondary purchases of the spectrum.

New Zealand

New Zealand was the first country to liberalize radio frequency management and use auctions to assign licenses. This was made possible with the adoption of far reaching reforms contained in the 1989 Radiocommunications Act whose main features follow:

– it enabled freely tradable spectrum management rights to be allocated using market mechanisms, with a relative neutrality on technology and usage;

– it required the registration of spectrum management rights according to a similar model used to register rights in real estate;

– it permitted the mortgage of spectrum management rights and it empowered spectrum management right holders to assign spectrum licenses under their management right[6].

There are three main players involved in spectrum management in New Zealand. These are:

– the Ministry of Economic Development where a Radio Spectrum Management Unit is placed;

– other Government entities, especially:

   - the Cabinet: all spectrum allocations are subject to Cabinet approval,

   - the Minister of Communications,

   - the Minister of Broadcasting,

   - the Ministry of Culture and Heritage,

   - the Puni Kokiri (The Ministry of Māori Development);

– the Commerce Commission (the competition law authority).

Inside the Ministry of Economic Development, the spectrum policy and management are performed by:

– the IT and Telecommunications Policy Group;

– the Radio Spectrum Policy and Planning Group;

– the Radio Spectrum Management Group.

---

6 Review of Radio Spectrum Policy in New Zealand (2005), Ministry of Economic Development.

Together, the last two form the spectrum management authority in New Zealand, referred to as the SMA.

The Minister of Communications and the Minister of Broadcasting stand at the top of the decision making process. Together and where appropriate in consultation with the Ministry of Māori Development and with Cabinet approval, they:

– approve all primary allocations;

– determine the spectrum which is to be reserved for social and cultural outcomes (broadcasting or telecommunications) or allocated to defense purposes.

The Minister of Communications is the primary authority as far as the Radiocommunications Act is concerned. The Commerce Commission is jointly the Competition law authority (under the 1986 Commerce Act) and the regulator of the telecommunications sector (under the 2001 Telecommunications Act). Merger rules and rules governing anticompetitive practices by dominant firms are administered by the Commerce Commission. Such rules apply to primary and secondary sales of the spectrum.

13.7. Is there an ideal structure for spectrum regulation?

As seen above, the institutional design of spectrum regulation is so diverse that it is difficult to come up with a simple framework describing all the alternative options.

Some cardinal distinctions are expressed through the patterns of spectrum regulation:

– competence on public service and/or commercial use?

– explicit political criteria or otherwise for the appointment of regulators?

– independent agency (quasi-autonomous) or governmental institution?

– split of competence on spectrum allocation (to services) and assignment (to licensees) or integrated responsibilities?

– competence on telecommunications and/or audiovisual?

– coverage of communications networks and/or content?

– electronic communications activities only or other network activities?

– regional dimensions or not?

Where Ministries are in charge of the spectrum, such as in Japan, China or India, a fully integrated competence exists but no independence as the ministries are parts of governments. On the contrary, in the USA for instance, public use is managed within the National Trade and Information Agency (NTIA), depending on the government, but an independent entity, the Federal Communication Commission (FCC), is in charge of the spectrum for commercial use. Its five commissioners are appointed along political lines (3 from the majority, 2 from the opposition). It is independent, however, in the sense that the commissioners, once in place, cannot be dismissed. It is "fully" convergent, in the sense that it is vertically integrated with regard to spectrum, performing both allocation and assignment, and also that it regulates telecommunications and audiovisual networks, as well as audiovisual content. Ofcom from the UK, as well as AGCom from Italy and BNetzA from Germany, are also convergent regulators, however, in Germany, the regulator relinquishes much of its competence on broadcasting spectrum to the federal states. France enjoys fully integrated horizontal allocation of the spectrum within ANFR and non-convergent wholesale allocation to the various uses, with the existence of the ARCEP, CSA, and other entities.

The previous examples have shown that various countries have made different choices for their regulation structures, in particular on radio spectrum management matters. These choices, eminently national, reflect different political constraints and different objectives to be reached. Does a privileged direction of evolution exist particularly in Europe? Is it urgent to reform?

The most recurrently discussed evolution, with the UK or Italy as examples, is whether it is appropriate to merge all the regulators dealing with commercial activities into a single convergent regulator, competent for both telecommunications and audio-visual, with appropriate spectrum management capabilities. Beyond technical motivations, such a convergence is in line with the evolution of all kinds of communication services towards a global sector of electronic communications.

The foreseeable convergence, in the long term, between audio-visual services and telecommunications was underlined by the 1998 European Commission Green Paper and there seems to be a natural justification to put together the regulators of these areas. The main objection, however, to a full regulatory convergence, is to assimilate a regulation of the audio-visual sector relating to contents and that of telecommunications which deals with networks. It is questionable whether network regulation, including wireless, implies the same competences as content regulation. For the former, the competences are mostly technical and economic. For the latter, societal, not to say political, considerations prevail. Mixing "economic" and "cultural" regulations is certainly delicate. Not only are the reference frameworks different but there is also a risk of bringing societal and political considerations too close to network regulation. Network regulation should be governed mostly by procompetitive considerations, with only a dose of regional planning and universal service added. It is also true, however, that the search for greater technical and economic synergies between these two sectors appears today as an enviable objective, likely to be better reached by a common regulator.

From a technical point of view, the development of convergent services, such as TV on mobile, and the concern of implementing "technologically neutral" regulations plead for merging regulators and regulations. Bringing together the spectrum parts dedicated to audio-visual and electronic communications is also a way to gain effectiveness and flexibility. It can facilitate swaps of frequency bands between the two domains without antagonistic lobbies playing one regulator against another. In line with the logic of convergence, it would be advisable that broadcasters pay a fee to use the spectrum, taking into account their respective public service obligations.

Another relevant reason for bringing together the regulators is an ever growing diversification of economic actors and products in a multimedia world. Gradually, the combination of services, operators and markets, with multiple mingled interests, makes it more difficult to single out situations and cases. Consequently, if specific reasons lead to preserve a sectoral regulator for electronic communications, it must be freed from too minute cases and able to look at them comprehensively.

Furthermore, the major importance of wireless services in the information society is becoming evident. Only radio provides the necessary permanent link connecting people. Consequently, the limited availability of the spectrum confers it an increasingly large economic value, to be only balanced by progress in technical and economic efficiency. Access to frequency bands becomes a major competitive factor for business and thus a convergent regulator needs to control such means of action.

Another reason to link the management of the spectrum to the general regulation of electronic communications is the prospect for a radical questioning of the traditional techniques of spectrum management in the area of commercial applications. We will see in the next chapter how the quest for more flexibility in spectrum usage and management may invent new techniques to challenge the traditional methods to allocate and assign frequencies and promote a unified management framework across uses. Such a long-term direction obviously results in recommending a weakening of the "technical" supervision on the spectrum. Simultaneously, the importance of financial considerations will grow with the value of spectrum and will justify an "economic" supervision much more coherent and comprehensive than today, entrusting the spectrum management authority with the regulation of a true frequency market.

13.8. Is a European regulator for the spectrum needed?

The electronic communications regulatory framework put in place in 2002 and progressively transposed and implemented until 2006, has created new committees and working groups in order to assist the European Commission in its task and facilitate a harmonized implementation of this framework. These structures are the Communications Committee, the Radio Spectrum Committee, the Radio Spectrum Policy Group and the European Regulators Group. The Cocom assists the Commission in exerting its executive powers under the new regulatory framework.

The Radio Spectrum Committee (RSC) assists the Commission in the development and adoption of technical provisions to ensure harmonized conditions for the availability and efficient use of radio spectrum. It is concerned with appropriate information available on spectrum usage. The RSC is part of the Commission.

The Radio Spectrum Policy Group (RSPG), on the other hand, is a consultative group distinct from the Commission. It is composed of Member State representatives, with the Commission acting as a secretariat. It has been established with the mission of advising the Commission on radio spectrum policy issues such as spectrum availability, harmonization and allocation, efficient use, methods for granting rights to use the spectrum, re-farming, relocation, valuation as well as protection of human health.

As we can see, Europe has got many structures which participate in spectrum management: the national regulators with a number of various components according to the States, CEPT, the European Commission and its ample "comitology", not to forget the "Committee of the Independent Regulators" which was created at the end of the 1990s and the "European Regulators Group", a coordinating structure which is progressively gaining political weight. This plurality illustrates the richness of Europe's political visions, expressed during the World Radio Conferences. It is also a source of inconsistencies, as could be witnessed with the UMTS license assignment story.

The question of a single European regulator recurs regularly, for example under the authority of the European Commission, the only supranational entity likely to hold truly European-wide legal capacities. This debate is currently (February 2007) reaching a climax.

The design of such an entity at EU level has not yet been clearly explored and its possible mandate remains vague. What model should the EU regulator refer to? Would it be the British or Italian model (convergent for commercial electronic communication services), the German model (ultra-convergent with a broad competence on public utilities), the French model (two levels, strategic at the EU's level with national operational "subsidiaries"), not to forget all others which have their legitimacy and interest. In addition the CEPT community is richer and more open than the European Union and can attract more countries in the orbit of the European theses.

In any case, it seems necessary, as a first step, to provoke essential subsidiarity between the activities which require a strong policy or a trans-national coordination, and those which must remain local for a better effectiveness. Another political question concerns the international competence of a possible European regulator with respect to the ITU which only recognizes the States.

An in-depth analysis would be essential before going towards the constitution of such an organization and it is not sure that it would lead to clear objectives without aiming to obtain well-defined advantages. The wisest direction today is to look at unsatisfied needs. A number of major subjects, already known and experienced, call for a truly political approach which leads to decisions to be applied by all countries within the framework of a constraining regulation and possibly defended jointly in front of international authorities such as the World Trade Organization. We will refer to the following examples of decisions or directions taken in recent years by the European Union and which had a strong capacity for drive, concerning the wireless services and the management of the spectrum:

– development of the GSM;

– development of the UMTS;

– launching of the GALILEO project of radiolocation by satellite;

– development of digital terrestrial television;

– R&TTE directive on the marketing of radio terminals;

– recommendation on the exposure of the public to radio waves.

On such subjects the contribution of the EU and the Commission is obviously very important. However, other subjects, more technical, can possibly be proposed:

– common European table for the planning of the spectrum;

– orbital-frequency resource management;

– economical principles for spectrum allocation and use;

– standard policy.

A partnership between CEPT and the European Union on such debates is essential, one bringing its technical skill, the other its political weight. This is presently undertaken.

In a somewhat dramatic move, in December 2006, the European Telecom Commissioner gave the national regulators a deadline to either propose the creation of a single telecommunications regulatory body for European Union – similar to the US Federal Communication Commission – or else surrender power to the Brussels-based EU executive body. Divergent applications of EU telecommunications rules and state controlled regulators made it increasingly hard for economic actors to operate across the 27 EU nations, insisted the Commissioner. To him, the only solution was to have telecom authorities centralized into a single, independent body. However, many national regulators fear that the Commission's push ignores the wide gaps between Europe's different telecom markets. Too much common regulation, too soon, would widen these gaps, not close them, was their argument. They said they were not against harmonization but the market contexts in all 27 European countries are very different. As an answer, the European regulators group pleaded for the Commission and the national authorities to "work more closely together" for a unified European telecom market while remaining within the current legal framework.

At this time, it is still undecided whether the creation of a single European communications regulator will actually take place.

Chapter 14

Major Current European and International Issues to Improve Spectrum Efficiency

Among many ideas to improve spectrum efficiency, the most straightforward is to refuse an ever growing "slicing" of frequency bands to allocate exclusive or priority resources to regulatory services, applications, operators and other resource applicants. On the contrary, the existing allocations and assignments should be regrouped as much as possible and their users incited to put in place appropriate techniques to cope with this new shared environment.

In this chapter, two major issues leading to such a regrouping are discussed: one concerning allocations to regulatory services, convergence, the other concerning assignments, collective bands. Both of them are actively studied by international bodies, notably in Europe.

14.1. Convergence

Major issues affecting spectrum management result from wireless services and technology convergences. With the publication by the European Commission of its "Green Paper on convergence of the telecommunications, media and information technology sectors and the implications for regulation"[1], convergence has been a topic of interest for several years. An important observation made in the Green Paper is the following quote:

---

1 Green Paper on the convergence of the telecommunications, media and information technology sectors and the implications for regulation – towards an approach for the information society /* COM/97/0623 final */.

> There is widespread agreement that convergence is occurring at the
> technological level. That is to say that digital technology now allows both
> traditional and new communication services – whether voice, data, sound or
> pictures – to be provided over many different networks.

This observation is still valid nowadays. However, during recent decades,
regulation has preferred to make a clear separation between the different regulatory
radio services (as specified by the ITU), e.g. mobile services, fixed services, mobile
satellite services, broadcasting services and so on. Now several examples become
apparent where new technologies blur the strict separation between these different
radio services (see Chapter 5). Convergence is on the way, notably between fixed
and mobile communications, as well as broadcasting services. The model adopted
for mobile services (standardization of technologies and high fees for frequencies)
often serves as a reference for spectrum management. Let us also remember that
satellite communications have their own convergence logic as fixed satellite service,
broadcasting satellite service and high density satellite networks are looking more
and more alike.

14.1.1. *Mobile – broadcasting convergence*

Mobile or personal TV is an example of convergence between mobile and
broadcasting services that currently gives rise to a great interest. Digital TV services
can be offered to mobile users based on different radio technologies. DVB-H (digital
video broadcasting to handheld terminals) is a standard derived from the regular
DVB standard initially targeted at fixed digital TV reception. Besides TV
broadcasting, both DVB-T and DVB-H can also support data services (datacasting).
With datacasting to mobile end user terminals, the service offered by DVB-H gets
closer to the 3G mobile networks. On the other side, the 3G networks can, in
addition to voice calls and data services, also support some kind of TV delivery to
mobile terminals.

What can be observed is that today mobile and broadcasting technologies are
complementary. The first can transmit low definition pictures to mobile sets, the
others high definition pictures to fixed sets. However, the technological
developments lead to convergence of both services in the close future.
Simultaneously, mobile and broadcasting operators are looking for a new
intermediate common market which is not fully identified but which could combine
features of their present offers. It is also noticeable that, going digital, the
engineering of broadcast networks is changing with many more stations transmitting
lower power carriers and placed on sites nearer to ground. Thus they may look more
and more like cellular mobile networks. Another interesting feature is that these new
digital broadcasting systems do not transmit fixed video channels but a high speed

digital frame which can be used much more fluently to carry any message, from HD video to local or personal data.

Looking at the spectrum management point of view, the allocation of the spectrum exclusively to one or the other radio service is thus getting less appropriate. In order to fully benefit from technological innovations that make the separation of mobile and broadcasting services partly disappear, this development should also be reflected in spectrum management.

14.1.2. Fixed – mobile convergence

The technological developments that are going on in broadband wireless access systems give rise to a comparable convergence between fixed and mobile services. The evolving standardization of such new networks (notably within IEEE 802.16) and the success of the WiMAX brand show this converging trend. WiMAX was initially a standard for fixed wireless access and therefore applied to the spectrum allocated for fixed services. More sophisticated transmission technologies are now becoming feasible that will not require the end user terminals to be fixed. With these enhanced radio transmission techniques introduced in the WiMAX standards, broadband radio communication services can be provided to nomadic and even mobile users. Then, for a telecommunications operator, the possibility of providing fixed, nomadic and mobile services using the same network infrastructure will offer economic benefits and enable new user applications in the concept of ubiquitous communications.

Chapter 5 has shown that many systems or devices are now introduced which are hybrids and cannot find a definite place between mobile and fixed services. SRD terminals, RLAN networks, now WiMAX networks fall into this intermediate category. Even satellite services, with high density systems, are concerned by the question. It should also be mentioned that the fixed long distance radio links, which were the very heart of fixed services and justified that a dedicated regulatory service be maintained to protect them against interference, with important bands allocated, are becoming obsolete and their bands reused for mobile services or fixed service for customer premises access. This evolution makes the individual frequency assignment and coordination process, adapted to such infrastructure links, less and less usable and useful.

This convergence of fixed and mobile services raises some spectrum management challenges since, in the current spectrum management framework, separate frequency bands are often allocated to fixed and mobile radio services. To provide maximum possibilities for technology innovations that support convergent fixed, nomadic and mobile services, it should be ensured that spectrum regulation will not create artificial obstacles. At this moment, CEPT is considering the possibility of introducing convergent fixed, nomadic and mobile systems in the 3.4 to 3.8 GHz band, referred to as the WiMAX frequency bands. Another example is the 2.6 GHz band that was designated in Europe for IMT-2000 systems. The promoters of WiMAX have argued that, for the sake of technology neutrality, WiMAX should be allowed in those bands as well.

14.1.3. *Wireless access platforms for electronic communication services (WAPECS)*

In line with the convergence of radio services, a coordinated European policy approach regarding spectrum availability for wireless access platforms for electronic communication services (WAPECS) is presently being discussed, in order to implement consistent spectrum approaches to consistent sets of wireless services.

The objective of WAPECS is "to harmonize frequency bands on a European basis in which a range of electronic communication networks and electronic communication services may be offered, on a technology and service neutral basis, provided that certain technical requirements to avoid interference are met, to ensure the effective and efficient use of the spectrum, and that authorization conditions do not distort competition".

The WAPECS concept uses the wording "electronic communications services" which includes IP access, multimedia, multicasting, interactive broadcasting and datacasting that could seamlessly be provided by a variety of radio communications technologies and networks. The WAPECS concept foresees further convergence of communication technologies and networks in the future, on common digital infrastructures. As depicted in Figure 14.1 the next generation of wireless communication systems could be a combination of fixed, mobile and broadcasting technologies as exist nowadays.

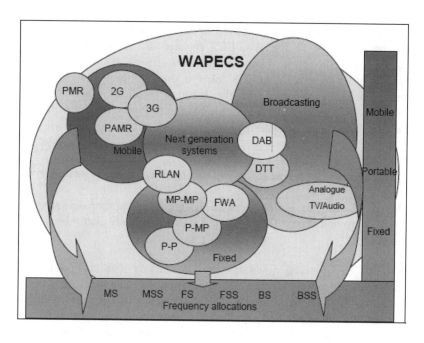

Abbreviations			
2G	Second generation mobile	MP-MP	Multipoint to Multipoint fixed links
3G	Third generation mobile	MS	Mobile Service
BS	Broadcasting Service	MSS	Mobile Satellite Service
BSS	Broadcasting Satellite Service	P-MP	Point to Multipoint fixed links
DAB	Digital Audio Broadcasting	P-P	Point to Point fixed links
DTT	Digital Terrestrial Television	PAMR	Public Access Mobile Radio
FS	Fixed Service	PMR	Professional (Private) Mobile Radio
FSS	Fixed Satellite Service	WAPEC S	Wireless Access Policy for Electronic Communications Services
FWA	Fixed Wireless Access	RLAN	Radio Local Area Networks

Figure 14.1. *WAPECS. Abbreviations. The wireless access policy for electronic communications services (WAPECS concept)*[2]

---

2 Wireless Access Platforms for Electronic Communications Services (WAPECS), "RSPG05-102 Final Opinion on WAPECS, 2005, www.rspg.europa.eu".

The frequency allocation to specific radio services (in the sense of the ITU Radio Regulations) does not fit the WAPECS concept. In WAPECS, the converged applications make no clear distinction between fixed, mobile or broadcasting services. In order to give WAPECS the full opportunity to develop, spectrum resources should be allocated for the broadest application possibilities. The allocation should be as broad as "electronic communication services" without stating if this is fixed, mobile or broadcasting use.

14.1.4. *Spectrum management issues imposed by convergence of radio services*

Is WAPECS credible, not only technically but as an appropriate instrument for all types of communications? Or is it utopia? It does not really matter because it asks good questions.

What can be observed nowadays are innovations that lead to a changing environment in which the clear distinction between the radio services as defined in the ITU Radio Regulations disappear. The convergence between fixed, mobile and broadcasting technologies and applications imposes challenges for spectrum management to remove some existing regulatory barriers in order to give these new developments sufficient possibilities to become successful.

In fact this means that not only should technological neutrality be introduced in spectrum management but also service neutrality. The first steps to adjust the spectrum allocation to technology and service neutrality are being undertaken at this time, but to fully meet the conditions for an unhindered introduction of emerging technologies offering converged services, a more flexible spectrum allocation will have to be introduced. Related general policy issues have also to be solved.

14.2. Collective use

The "commons" approach was made very popular in the early 2000s, as the advent of WiFi seemed to introduce a far-reaching new model of spectrum management based on license-free access to frequencies. Considering the perspectives offered by this innovative approach, the possible extent of "commons", or more adequately called "collective use" bands, initiated comprehensive studies[3].

---

3 Study on Legal, Economic and Technical Aspects of "Collective Use" of Spectrum in the European Community, Mott MacDonald Ltd *et al.*, November 2006.

14.2.1. *Types of spectrum collective uses*

The term "collective use of spectrum" includes the well established use of the spectrum on a license exempt basis as well as more recently developed sharing concepts such as underlay and overlay, described below. Collective use spectrum can be application specific, technology specific, or neither of these. Each approach has its own merits depending on the applications that are expected to use the spectrum, in particular for the service quality and the likely interference environment. Some are listed here.

License exempt (commons) – non-specific applications: no individual authorization or coordination is required and no fee payable for using the spectrum. Access is only regulated by compliance to pre-defined regulatory conditions (typically specified in the national frequency allocation table and national legislation) which may be based on EU or CEPT harmonized specifications. Any application is permitted so long as the regulatory conditions are respected. The corresponding applications are typically low power and short range.

License exempt (commons) – specific applications: no individual authorization or coordination is required and no fee payable for using the spectrum. The equipment must comply with specific standards, either harmonized standards or national interface standards which relate to specific applications. They are typically low power, short range devices but also, for example, leisure personal radio terminals, such as CB radios or walkie-talkies.

Light licensing – few restrictions: registration or notification is required. No limits are put on the number of users but the usage may be application specific. Typically, light licensing authorizes a transmission of greater power than license exempt applications. A small fee may be payable to cover the costs of the registration or notification scheme. Light licensing typically applies in situations with no immediate concern about interference but where there may be particular service regulations with a need to make changes to these regulations and use of the spectrum, in the future: thus there is a need to maintain a record of users. For example, some European countries allow the use of the 5.8 GHz band for fixed wireless access services on a light licensing basis without the need to apply for an exclusive license or right of use. Short distance radio maritime terminals or amateur services may also fall into this category.

Light licensing – with restrictions: registration or notification is required and there are limits on the number of users with requirements for coordination between them. Usage may be application specific and typically allows greater power than license exempt services. A small fee may be payable to cover the costs of the registration or notification scheme. Recent examples include a registration scheme proposed in the USA for use of the 3,650-3,700 MHz band on a collective basis for fixed wireless access, where the risk of interference is mitigated by technical means and where licensees are mutually obliged "to cooperate and avoid harmful interference to one another". In the same way, the UK regulator, Ofcom, recently awarded through auctions 12 low power concurrent rights of use for the frequencies 1,781.7-1,785 MHz paired with 1,876.7-1,880 MHz. Licensees are expected to coordinate their use of the spectrum to avoid harmful interference.

Private commons: an individual right of use is required from a licensee who may sub-let the access to the spectrum to third parties on an unlicensed basis without the need for coordination, so long as pre-defined regulatory conditions are accepted. Responsibility for avoiding interference with users outside the licensed spectrum band rests on the licensee. In the USA, the Federal communications commission (FCC) recently introduced rules permitting spectrum leasing under which a licensee may rent a block of the spectrum to create private commons for use by thousands or even millions of users. The FCC speculated that this type of private commons may interest innovative equipment vendors to roll out new services such as a private WiFi business with a higher service quality than ordinary WiFi systems as installed today in existing license exempt bands.

Experimental commons: experimental licenses are intended for use on an experimental basis for predefined and limited periods of time. License, registration or notification depend on the specific allocation but there are generally no limits on the number of users and there may be no restriction on applications. Operation is provided on a non-interference, non-protected basis, and operational constraints may apply (e.g. prohibition on provision of third party services). Pre-defined technical constraints may apply or may be specifically negotiated. A small fee may be payable to cover the costs of the registration or notification scheme.

Underlay: underlay technologies operate in the spectrum that is already used for other licensed or license exempt uses but at a very low power level which does not impair them. This allows the underlay use to share the spectrum with other traditional services. Underlay use is not licensed. Ultra wideband (UWB) is an example of an underlay technology.

Overlay: an overlay approach permits higher powers that could cause interference to existing users but overcomes this risk by only permitting transmissions at times or locations where the spectrum is not currently used. This can be achieved either by technology (e.g. cognitive radio) or by regulatory means (e.g. only permitting use in particular geographic areas). Of course here only unlicensed overlay uses are considered.

These different categories may also be combined in a given frequency band. For example, a scheme, called "private spectrum parks", is under consideration for wireless access services in Australia. This approach combines licensed, registered and license exempt uses in the same frequency range but applying to separate geographic areas. In this case, licensed use could apply in the main cities, registration (with coordination) apply in slightly populated rural areas and license exempt use apply in remote areas with very low population density.

14.2.2. *Protection against interference*

Unlike exclusive use of the spectrum, notably with frequency assignment and registration, collective use does not confer on their users any protection from other spectrum users (licensed or exempt) who are operating legally in conformance with relevant technical and regulatory parameters. They are clearly in a secondary status beside primary users and services, if any, and cannot benefit any priority or protection from other users, at their level. For some types of collective use where the quality of service is of particular importance (such as professional wireless microphones or wireless medical devices), the probability of interference can be minimized by allocating application specific bands and applying additional technical constraints (e.g. power or duty cycle limits). In other cases there is a presumption that the user can tolerate a reasonable level of interference, at least statistically, or that appropriate interference mitigation techniques will be adopted.

As a summary, the following hierarchy can be defined for the interference protection that a user can expect:

– full protection: exclusive right of use is required. Legal recourse is provided in case of harmful interference;

– partial protection: collective use is restricted to specific applications and technologies. There is no legal recourse against interference from other legitimate spectrum users but users and applications may sometimes be known to facilitate bilateral and informal coordination;

– no protection: collective use, the spectrum is open to all applications and technologies compliant with basic technical parameters; no legal recourse is possible against interference from other legitimate spectrum users who are anonymous.

It can be thought that spectrum pricing should be adapted to such different situations with a market-oriented policy for fully protected bands, limited administrative pricing for partly protected bands and free use for unprotected bands.

14.2.3. *Collective spectrum in Europe: the present situation*

There is a good level of harmonization of unlicensed frequency bands in Europe reflected in the ERC recommendation 70-03. No technology is specified in this CEPT document, some applications are however mentioned in specific appendices in order to avoid interferences. This 70-03 recommendation is currently used in 47 countries around the world (Europe, New Zealand and Australia, etc.).

Historically, the approach to harmonize collective spectrum use in Europe has been mainly on the basis of recommendations or decisions issued by CEPT. There are some exceptions, such as the spectrum allocation to DECT cordless phones, which is mandated by a European Commission (EC) directive.

CEPT recommendation effects are limited: there is no obligation on individual Member States to implement them. However, decisions have a greater weight in that, once Member States have committed to implement them, they are obliged to do so, usually by means of transposition into their national legislation or incorporation into their national frequency allocation table. However, there is no obligation on Member States to commit to CEPT decisions.

Collective use spectrum is mainly allocated to short range devices (SRD), which is the generic term for a number of applications and technologies. SRD equipment complies with the EU RTTE directive (termed Class 1 products) and is covered by another CEPT decision covering license exempt use.

Since the introduction of the new EU regulatory framework, a closer working relationship between the European Commission and CEPT has been established and there has been a move to use Commission decisions to back key harmonization measures proposed by CEPT. Typically, the European Commission radio spectrum committee identifies a need for harmonization. Then it instructs the CEPT ECC to undertake necessary market or technical studies for the harmonization process and to develop a draft EC decision based on ECC deliberations. Later, an EC decision is transmitted to the European Parliament for ratification following a process of internal (EC) and external public consultation. The process is illustrated in Figure 14.2 below.

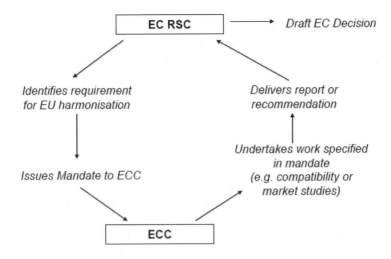

Figure 14.2. *Harmonization process involving EC and ECC. Source: study on legal, economic and technical aspects of "collective use" of spectrum in the European Community*

An example of such a process has taken place with regard to short range devices (SRD). A draft Commission decision on the harmonization of the radio spectrum for use by SRDs was issued for public consultation on 8th August 2006. The drafted document requires Member States to make available, on a non-exclusive, non-interference and non-protected basis, a set of frequency bands for short range devices, subject to specific conditions. The decision would be regularly revised to be adapted to technical progress.

Appendices to the draft decision reflect the content of a number of appendices of the existing CEPT ERC recommendation 70-03, mainly those which have already been implemented in all EU Member States under the existing voluntary arrangements. By inclusion in an EC decision, the specified bands and applications would become mandatory in all Member States. The current draft decision does not address bands and applications that have not yet been implemented throughout the EU and are thus more likely to be of concern to manufacturers and potential users. In this sense, the EU's current role in this area appears to be largely a reactive one: it strengthens the status of already existing ECC decisions on harmonized frequencies, technical characteristics, exemption from individual licensing and free carriage and use of several radio systems.

ERC Recommendation 70-03 (referred to as ERC 70-03) provides a consolidated set of harmonized frequency bands and applications for short range devices in the CEPT area. It is a lengthy and complex document, concerning more than 60 frequency bands, with 13 appendices covering different types of SRD applications. The extent to which each of the recommended bands is actually harmonized among CEPT countries varies considerably. Information on the extent of harmonization (i.e. which countries have implemented or committed to implement each frequency allocation identified in the document) is also provided. The appendices cover 12 specific applications and a single generic category of "non-specific devices". Most of the application specific appendices relate to applications where safety or security may be an issue, such as transport (rail and road), alarm systems, medical devices or movement detection. Interference to such applications could have serious consequences, but the characteristics of the systems themselves (e.g. short duty cycle, occasional intermittent operation, very short range) mean that interference between these devices is unlikely to happen. For these applications, unconstrained sharing with other types of SRD may present an unacceptable risk of failure and justifies specific frequency allocations. Some other applications, whilst not having the same safety or security issues, may also be unsuitable for unconstrained sharing with other devices. For example, interference to radio frequency terminals of data systems could produce significant disruption in retail and warehouse environments, or interference to wireless microphones could seriously disrupt major public events where these devices are used. All of these various constraints notwithstanding, applications such as wireless data transmission and wireless audio should in principle coexist well with other SRD applications so long as appropriate sharing protocols are deployed.

14.2.4. *Challenges*

European regulators are facing a number of challenges related to the collective use of the spectrum. A major difficulty for a European policy of collective use bands, notably dedicated to SRD, is to control the trans-border flow of non-compliant devices which are bought outside and operated without concern for local regulations. Having entered a European country, they are almost impossible to locate and monitor. Low power FM microtransmitters and UWB equipment are examples of these new challenges associated with differences in the regulatory frameworks between Europe and the USA.

Low power FM microtransmitters

This example relates to the illegal importation of low power FM microtransmitters used to transmit music from portable devices to FM radio receivers. These devices became widely available in the USA and Far East, where they could be operated legally under existing provisions catering for ultra low power devices. As no such provisions existed in Europe, use of these devices was illegal but many thousands were imported into the EU, both by returning travelers and on a commercial basis. Studies were undertaken to derive appropriate limits for interference mitigation and steps were taken to include provision for the devices in Recommendation 70-03 and the related ETSI standards. However, many countries have still to implement this provision.

UWB equipment

In the same way it will be very difficult for European regulators to control the use of UWB equipment produced in Asia for the US market. These devices (wireless USB devices for instance) operate on frequency bands not open to unlicensed equipment in Europe, as the spectrum masks defined by the FCC and the ones defined by CEPT are not the same.

14.2.5. *Impact of an extension of unlicensed bands on spectrum planning*

A study on the legal, economic and technical aspects of "collective use" of the spectrum in the European Community issued recommendations on new frequency bands for SRD which should be considered as potential future candidates for collective spectrum use.

Frequency band	
47-68 MHz	This will no longer be required for broadcasting following digital switchover and is unsuitable for other licensed use due to the risk of interference arising from abnormal long distance propagation effects. This band could be attractive for longer range, higher powered collective use applications.
862-872 MHz	This band should be made available on an application neutral basis, subject to the use of spread spectrum technology to minimize the risk of interference to existing users of the band. This would provide a European alternative to the established and successful 900 MHz ISM band in North America. This extended band would make use of the spectrum on either side of the existing 863-870 MHz collective use band that is currently allocated to other services in some Member States but is basically underused.
915-917 MHz	Consideration should be given to the introduction of RFID interrogation systems in this band, to cater for growing demand for these devices. Compatibility studies should be undertaken to investigate the potential impact of such systems in a variety of deployment scenarios on GSM and UMTS base station receivers operating immediately below 915 MHz.

40.5-42.5 GHz	This band, currently harmonized for multimedia wireless systems, should be considered as an early potential candidate for collective use, in view of the absence of any current interest in the use of this spectrum for licensed system deployment. In the longer term, consideration should be given to make most of the spectrum above 40 GHz available for collective use, with the possible exception of those bands already identified for specific licensed applications, such as the fixed link band at 55 GHz, or where there are other services such as radio astronomy that may require protection at certain locations.

Table 14.1. *Potential frequency bands for collective spectrum use. Source: Study on Legal, Economic and Technical Aspects of "Collective Use" of Spectrum in the European Community*

An extension of the unlicensed bands may reduce the share of the licensed ones in frequency bands below 30 GHz. The balance for bands between licensed mobile services and unlicensed WLANs is sensitive and should be carefully adjusted. This is especially true below 6 GHz as the frequency bands above 6 GHz are not very attractive for mobile operators, due to propagation limitations. Unfortunately, an in-depth analysis of collective use possibilities demonstrates that what can be reasonably proposed is limited. However, the introduction of dual mode mobile terminals which add WiFi capability to GSM terminals is interesting. If a significant share of voice calls is made on WiFi instead of the cellular networks in the coming years, this could decrease, in some places, the traffic congestion in the licensed bands of the mobile operators. A limited but non-negligible increase of service quality could be achieved by combining licensed and unlicensed frequency bands.

14.2.6. *An example of introducing collective use in radar frequency bands*

This example is interesting as it shows that it is possible after detailed technical investigation to find new frequency bands for collective use. Here, the availability of technological developments enabled the introduction of WLANs in a frequency band allocated to radar[4].

Studies on developing RLAN systems in the 5 GHz band began in early 1991 within CEPT following a request from ETSI, which had started its standardization for the HIPERLAN system. It was a recognition that the 2.45 GHz band already identified for RLAN, with its limited capacity (83 MHz) and the severe sharing

---

4 Enhancing harmonization and introducing flexibility in the spectrum regulatory framework. Electronic Communications Committee within the European Conference of Postal and Telecommunications Administrations – ECC report 80 – Oulu, March 2006. Implementation of an innovative spectrum management technique: the dynamic frequency selection and the RLANs at 5 GHz.

environment (ISM, Short range devices), would create a risk of impairing the development of RLAN, although it was at that time mainly a concept.

CEPT designated in 1992 the band 5,150-5,250 MHz for the HIPERLAN, taking into account the sharing constraints with other services. An "optional" 50 MHz in the band 5,250-5,300 MHz, to be shared with radar systems, was also proposed on a national basis.

Later, a concept similar to HIPERLAN was developed in the USA under the name "U-NII". This led to the identification by the FCC in 1996 of the band 5,150-5,350 MHz and 5,725-5,825 MHz for such applications, significantly broader than that designated by the CEPT. Also, 200 MHz of this spectrum was allowed for outdoor environment.

Therefore, ETSI came back to CEPT asking for a similar amount of spectrum, justified by the need for a sufficient number of channels enabling quasi-cellular deployment, and for some spectrum authorized in outdoor environment. CEPT initiated additional studies but faced two difficulties. Firstly, the band 5,725-5,825 MHz was allocated in Region 1 to the fixed satellite service, making this band unsuitable for outdoor applications. Secondly, the whole band 5,250-5,825 MHz was used for various types of radar and it was demonstrated that a large population of RLANs would have a high potential for interfering with such radars.

In this context, the idea of applying dynamic frequency selection (DFS) to ensure the compatibility with radars was successfully developed in 1998 and 1999 within CEPT. The DFS concept (see Chapter 10, section 10.4) was not really new: it was often used to enable coexistence between digital equipment in the same system (e.g., in the DECT system). However, it had been the first time that it had been used as a mechanism for enabling sharing between systems as different as RLAN and military radar.

In practice, the definition of exact DFS characteristics required much work to define the level at which a radar signal will be detected and how the RLAN should behave when detecting such a signal. In parallel, CEPT proposed at WRC-2000 that RLANs in the 5 GHz band should be included in the WRC-03 agenda in order to get an additional allocation for the mobile service to enable RLANs to use this band which was also partly allocated to space services (RLANs are considered by the Regulations as mobile services; see Chapter 5, section 5.1) After long discussions with the USA, in particular, the band 5,470-5,725 MHz was finally allocated to mobile services on a primary basis, subject to an obligation for mobile stations to use DFS.

This illustrates the fact that CEPT was able to identify and implement successfully a new sharing technique enabling the designation of a very large part of the spectrum (455 MHz) for RLANs in the 5 GHz band. This technique was subsequently generalized to other parts of the world through WRC-2003 decisions. Compatibility studies organized by administrations to arbitrate between incumbents and new entrants (operators or manufacturers) have enabled a modification to the allocation framework in a timely and successful manner, with the opening of an important band to unlicensed systems.

Chapter 15

Regimes of Radio Spectrum Management:
A Synthetic View

Wireless services will undergo great expansion in the coming decades. The generally accepted view is that this will create an increasing need for radio spectrum. Major technological changes are also under way, which might help to improve its efficient use but also provide new and cleverer management methods[1].

The present chapter deals mainly with business-oriented and competitive services to customers for telecommunications and broadcasting which stay at the very center of concerns since mobile communications surged as mass markets. However, the ideas may apply to some other radio services.

The debate on adequate, future-oriented, spectrum management is currently reaching a critical point. In addition to technical advances to achieve a more flexible spectrum sharing (see Chapter 14), a trend towards market mechanisms (auctions of spectrum property rights and trading) was initiated in New Zealand, then in the USA in 1993, and expanded in a number of European and Asian countries. It was consistently formalized in 2002 by the FCC Spectrum Task Force Report in the

1 This chapter is expanded from an article published in Communications & Strategies in April 2007. It has strongly benefited from the research being carried out within the European Commission specific support action SPORT VIEWS (Spectrum Policies and Radio Technologies for Viable Wireless Services, Contract No. 027297). Observations and contributions have been received from Frédéric Pujol, Marvin Sirbu and anonymous referees. However, the authors carry sole responsibility for the views expressed in this chapter. They do not in particular necessarily reflect those of the European Commission or the SPORT VIEWS consortium partners.

USA[2] and the Martin Cave Report in the UK[3]. The two reports[4] translated into a comprehensive, market-oriented spectrum management framework have been used as an underlying reference for subsequent policy initiatives in Europe since that time.

It seems, however, that this trend is encountering delays and limitations where it is already implemented or favorably considered, and facing fierce opposition in some territories it has not conquered. It might consequently be useful to sort out the alternatives, list the arguments exchanged, look into the prospects offered by ongoing technological wireless developments, and pave the way for possible transition paths.

A few preliminary statements and general background observations might help to set the scene before proceeding:

– spectrum is a limited resource, as it is often said. As a methodological caveat, however, it does not directly follow that spectrum is *scarce*, especially as continuous and significant improvements in efficiency in spectrum usage are under way, as illustrated in many research programs. It is rather a permanently constrained resource (see Chapter 4, section 4.6);

– however, and somewhat paradoxically, spectrum is under-utilized and better management is needed to help at least partially overcome use limitations;

– spectrum limited availability means higher costs. License fees have reached in some cases amounts comparable to network equipment investments. Nowadays, the cost of spectrum may be a major factor in determining the business plan of new systems. The diffusion of services would hugely benefit from all technologies and management methods allowing for lower spectrum costs;

– high fees or prices can be justified, at a specific time, as in the case of oil for instance, as a mechanism for saving reserves or to finance exploration activities, but they justify that a greater effort be made for a more efficient usage;

– the granting of exclusive or partly exclusive rights on limited resources for market-oriented activities must be subject to competition and conveniently priced by market-oriented mechanisms.

------

2 Federal Communications Commission Spectrum Policy Task Force Report, November 2002 http://www.fcc.gov/sptf.
3 *Review of Radio Spectrum Management, An Independent review for the Department of Trade and Industry and HM Treasury*, by Professor Martin Cave, March 2002, http://www.spectrumreview.radio.gov.uk.
4 For an overview of spectrum management methods and the international experience: J. Scott Marcus, L. Nett, M. Scanlan, U. Stumpf, M. Cave, G. Pogorel. *Towards more Flexible Spectrum Regulation*, WIK-BundesNetzAgentur, December 2005.

Different radio spectrum management regimes are defined according to four types of questions:

– Should frequencies be allocated according to a harmonized plan?

– Should the allowed technologies be standardized?

– Should spectrum usage rights be exclusive or collective?

– Should usage rights be assigned through market mechanisms, administrative procedures or hybrid procedures?

Facing such questions, we intend to:

– explore the widest range of choices available to regulators and industry in establishing a radio spectrum management policy;

– organize and clarify the expanded set of alternatives to be considered;

– list the criteria whereby the necessary choices and decisions can be made.

We propose a balanced set of decision criteria for each question. Thus, nine regimes result from the combined answers. This taxonomy illustrates the possible rationales for a diversity of regimes broader than the usually exposed standard trilogy of Command and Control, Market and Commons. The nine regimes can also be considered as a map to be navigated in order to fit with institutional and technological transitions. This allows decision makers to come up with informed choices using all the technical information available and based on definite criteria and a rigorous methodology.

15.1. Definitions: four dimensions of spectrum management

A spectrum management regime is built on four dimensions that have to be successively analyzed with alternative approaches:[5]

– Service frequency allocation: spectrum harmonization or neutrality.

– Technology: technology standardization or neutrality.

– Usage rights definition: alternative regimes.

– Assignment modes of spectrum usage rights.

---

5 Johannes M. Bauer, *A Comparative Analysis of Spectrum Management Regimes*, Quello Working Papers, 2002.

15.1.1. *Frequency allocation to services: spectrum harmonization or neutrality?*

Spectrum harmonization is understood here as allocating a frequency band or set of frequency bands to a particular application or a category of services (regulatory services), including variants between regions or countries. It depends on regulatory provisions set at international (ITU), regional (Europe, America, Asia) and national levels and imposes limitations on service neutrality.

Harmonization does not necessarily have to be implemented all over the spectrum. There can be harmonized bands, if justified, and non-harmonized bands elsewhere in the spectrum. However, nowadays, the Radio Regulations allocation tables cover the whole usable spectrum, leaving no band without a service allocation, at least theoretically.

There is still a question of whether the traditional distinction between regulatory services remains appropriate with convergence (see Chapter 14, section 14.1) and whether more generic services can be designated.

15.1.2. *Technology: standardization or neutrality?*

Standardization is understood as designating a particular technology or set of technologies to implement an application or a service category. It aims to ensure that the equipment used meets technical requirements, specified in technical product standards, or specifications, in order to provide market advantages for a better coexistence or interoperability, cross-border roaming, economies of scale, etc. Standards can be determined by public bodies such as ETSI or the market (industry-led), and then mandated by regulation or not. To simplify, it is assumed that standardization can only reasonably occur in a harmonized context.

It is important to draw a distinction between harmonization and standardization, as understood here. Harmonization of frequency bands concerns the frequency allocation level as standardization takes place at a technological level. There can be harmonization with or without standardization. Of course there are obvious relationships between these two, as technical standards have to fit with the physical qualities of frequencies. Yet harmonization and standardization do not necessarily go hand in hand. Various technical standards or even non-standardized technologies can possibly be used in harmonized bands. Although this distinction is needed, the confusion can be found in some studies on spectrum policy[6].

---

6 Booz-Allen, *Study for the UMTS Forum*, 2006.

An intermediate category does exist between harmonization and standardization which is channel planning. A band can be divided into standard channels which are identical sub-bands whose parameters, notably their width, are defined. A channel is designed to locally accommodate a radio service, and is related to some standards, with neutrality. As an example an 8 MHz TV channel, in Europe, fits with traditional analog TV or DVB-T digital TV (see Chapter 2, section 2.1). Here, channel organization will be considered as a part of standardization.

15.1.3. *Usage rights definition*

Property rights have been widely heralded as a major factor of economic and social dynamic evolution. In the spectrum context, the following categories of usage rights can be defined:

– exclusive property rights (without easement, but in conformance with very general purpose spectrum regulations);

– property rights with easement. Easements are intended here not as temporary, but permanent disaggregation of property rights to allow for more efficient use;

– provisions are made to ease band sharing, such as overlay, underlay and dynamic frequency selection (DFS) (see Chapter 14, section 14.2). As an example, DFS can be requested in a restrictive sense as the possibility of shifting between a set of predetermined harmonized bands or, more extensively, as the possibility of shifting across large areas of the spectrum. It is compatible with all non-exclusive property-based regimes. This perspective of mitigating property rights to benefit from technological evolutions has been extensively explored by Martin Cave[7];

– collective use, the third type of usage rights, became popular as "commons" in the late 1990s with the advent of WiFi and was promoted as a far-reaching, future-oriented model. The possible extent of collective use and, conversely, of property rights is discussed in Chapter 14, section 14.2.

---

7 Cave M., *New Spectrum-using Technologies and the Future of Spectrum Management: A European Policy Perspective*, Ofcom, May 2006.

15.1.4. *Assignment modes of spectrum usage rights*

Some frequencies or channels may be assigned to operators by purely administrative methods, without competition, when no scarcity is observed. Thus, the basic rule is "first come, first served". This traditional method remains appropriate and internationally accepted for a number of applications such as satellite networks (see Chapter 8).

Here will be considered the procedures which can be used when only a limited resource is available for a market-oriented business with competitors. There are two main categories of usage right assignment modes:

– comparative administrative procedures include:

- pure administrative procedures such as beauty contests. They may include an administered incentive pricing[8],

- hybrid modes such as administrative procedures with a bidding price as part of a weighted multi-criteria formula. There is a price component but the license remains under administrative control;

– auctions resulting in exclusive property rights, which represent the quintessential market solution for the assignment of spectrum usage rights. Trading is a complement to this approach for secondary markets.

Assignment modes have been the subject of numerous studies[9]. The table below summarizes the overall alternatives arranged in a four-step decision tree designing nine spectrum management regimes.

---

8 International Telecommunication Union, Telecommunication Development Bureau, F. Château, Ch. Picory, Tariff policies, tariff models and methods of determining the cost of national telecommunication service, Document 1/129-E, 12 July 2000.
9 See above FCC and Martin Cave, both 2002.

15.2. Choosing a spectrum management regime

STEP 1 Frequency allocation: harmonization or not	STEP 2 Technologies standardization or not	STEP 3 Usage rights	STEP 4 Spectrum assignment mode	Spectrum management regime #
Harmonized spectrum (no service neutrality)	Standardization (no technology neutrality)	Property rights: exclusive	a/ Administrative assignment Procedure/hybrid	1a Standard command and control (CC) 1b Technology control/property rights (PR) market
		Property rights: with easements	b/ Auctions/trading	2a Mitigated CC with easements 2b Technical CC+ mitigated market
		Collective use	License-exempt	3 CC Collective
	Techno neutrality	Property rights: exclusive	a/Administrative assignment Procedure/hybrid	4a Technology neutrality in CC context 4b Harmonized neutrality
		Property rights: with easements	b/ Auctions/trading	5a Controlled neutrality 5b Harmonized neutrality plus
		Collective use	License-exempt	6 Standard "commons" regime
Service neutrality NO Harmonization	NO Standardization	Property rights: exclusive	a/Administrative assignment Procedure/hybrid	7a Administered neutrality 7b Pure market regime: libertarian
		Property rights: with easements	b/ Auctions/trading	8a Technology neutrality/administered semi-PR Market 8b Mitigated market regime: semi-libertarian
		Collective use	License-exempt	9 California dream

Table 15.1. *Nine spectrum management regimes: a four-step decision guide*

The standard trilogy is present in this table. We recognize *Regime 1a* as the traditional command and control model which is still much used. *Regime 7b* can be qualified as a full property market-based regime, combining flexible frequency allocation and technical choice. *Regime 6* is the "commons" model.

What this table illustrates, however, is a diversity of regimes broader than usually described. Before looking into the basis on which each regime could be pragmatically justified, let us provide some brief descriptions. Regime 1b, for instance, combines a traditional command and control scheme for frequency allocation and technology, and auctions for property rights assignment. Harmonization is thus compatible with some market mechanisms for assignment, as seen in the UMTS case in Europe. This represents some kind of limited or harmonized neutrality: the application is fully constrained for its technical aspects but the business activity may be fully market-oriented.

Hybridization can also occur between comparative administrative procedures and auctions, as in *Regime 4a technology neutrality in a command and control context*. An example is the 2006 "WiMAX" authorization procedure in France. It combined qualitative elements submitted to administrative evaluation, like the contribution to regional development, with financial bids. It could also be presented as auctions with room for negotiation and mediation by the regulator. The frequency bands were harmonized, being allocated to the fixed service, but the technology choices were not standardized.

Regime 5b harmonized neutrality plus looks like an interesting combination of harmonized frequencies, technical neutrality and easement on property rights acquired through auctions, thus accommodating some operator preferences, as well as possibilities of spectrum sharing and dynamic frequency selection.

Not all regimes are representative of realistic alternatives, but homage should be paid to *Regime 9, California dream*, which embodies the vision of large radio spectrum commons, open to any use, assuming that technologies exist to support it.

We have shown that neutrality and efficiency in spectrum management have to be considered at various levels and can be combined in a variety of ways. Let us now expose the iteration of pragmatic considerations on four successive levels of analysis, leading to the choice of a particular spectrum regime.

15.3. Deciding on spectrum management regimes: a four-step process

Efficiency is often quoted in relation to neutrality. CEPT ECC Report 80 defines neutrality[10] as:

> Increasing the ability of the spectrum regulatory framework to facilitate and adapt, in a timely manner, to user requirements and technological innovation by reducing constraints on the use of spectrum and barriers to access spectrum.

Neutrality is easily accommodated within the proposed analytical taxonomy, as it takes place at all four decision levels: frequency allocation, technology, usage rights and assignment procedures. We now have the possibility of looking at spectrum efficiency in relation to this comprehensive and articulated framework to gain a better understanding of what is at stake in the present debate. Let us look at the criteria that decision makers can resort to when going through the four steps which make up spectrum policy choices.

On the two issues of harmonization and standardization, it should be noted that the existing literature provides neither positive nor negative compelling evidence on their overall necessity and superiority to non-harmonization and non-standardization. Moreover, the existing literature does not provide either a locally applicable toolbox of criteria to make the proper choices easily. The evidence itself is not yet conclusive. For instance, in the very popular area of mobile communication services, the jury is still out on the outcome of the confrontation between the GSM-UMTS standardization line of action in Europe and the agnostic approach adopted by the USA and Korea. Industry associations, however, at least in Europe, emphasize the benefits of harmonization and standardization in two recent studies (GSMA and UMTS Forum, 2006).

Let us examine some critical questions to be answered when making choices of harmonization versus non-harmonization and of standardization versus non-standardization.

Step 1: harmonization?

Harmonization is meant to minimize interferences, reduce cross-border coordination requirements and ensure roaming facilities. The benefits for consumers result from lower network planning expenses and lower prices of devices. The costs of harmonization are the inefficiency costs incurred from local or overall suboptimal usage of the spectrum resource, administrative costs and slower innovation.

---

10 CEPT-ERO Report 80: *Enhancing Harmonization and Increasing Flexibility in Spectrum Management*, www.ero.dk, March 2006.

The question to be answered by decision makers is: how much does the lack of, or only partial, harmonization (as in the case of GSM bands in Europe and the USA) impact the cost of network equipment, terminals and services? A common strategy to combine advantages may be to design a harmonized core band which is available everywhere and extension bands which may be different from one country to another. A variant is to design standards and equipment with a wide tuning range but to use a restricted and different part of it in every country (the case with WiFi, as an example). On the contrary, a bad strategy would probably be to choose a fully harmonized band even if it was too narrow for true competition, creating an artificial scarcity.

Notwithstanding this debate, as a first step to implement any new service, international regulators generally try to achieve worldwide or regional spectrum harmonization as a fundamental objective. As a second step, they try to cope with actual local constraints.

Step 2: standardization?

It is generally accepted that there is a direct negative relationship between production scale and the cost of manufactured products. Standardization leading to an increase in scale is intended to lower costs. The argument against standardization (government or even industry-led) is that it creates a lock-in, which slows or even precludes the introduction of innovative unexpected technologies.

At the highest conceptual level, dynamic efficiency must prevail upon static considerations: innovation being largely unpredictable, government-led standardization would have to be an exception, based on strong arguments, and industry-led standardization to be carefully monitored to avoid the establishment of barriers to entry. Eventually, the trade-off is between lower costs made possible by economies of scale and potential barriers to entry for innovative new technologies.

If it can be safely assumed that the market demand and technical progress for network equipment and terminals in a significant period (10-15 years) can be anticipated and accommodated within designated standards or standard categories, the benefits of standardization apply. If too much uncertainty regarding future service needs or technologies exists, or if a careful examination of developments in the labs leads us to assume there is a risk of major disruptive changes, avoiding standardization is the safe bet.

The past experience of modern cellular mobile telephony in Europe is interesting. The first generation (analog) was non-standardized with various standards, such as NMT or TACS, which created a dynamic market for voice on mobiles in different countries. The second generation, GSM, providing nearly the same basic service as the first one, was standardized and met an extraordinary success throughout the world. The third generation, UMTS, also standardized, seems to have some difficulties in achieving its objectives. Was it not preferable to leave the technological choices open to find the most appropriate way to satisfy the fully new market of mobile multimedia contents?

Questions about standardization are: how much is lost as extra costs for consumers if terminals combine two or more standards? Is our view of the technology and market evolutions for the next 10-15 years accurate enough to aim at the benefits of standardization or is there a risk we might miss valuable opportunities or make major mistakes?

Step 3: what type of usage rights?

Exclusive property rights

Property rights, in the area of the spectrum, are favorably described as fostering an efficient use, allowing players to control the resource and, when combined with trading, introducing an element of smooth and efficient neutrality in accordance with economic optimality criteria. Many advocate that they must be exclusive (without easement) to confer the licensees the benefit of a "clean spectrum", free of interferences.

The costs and potential risks relate to abuses of a privilege. An exclusive right may facilitate the creation of entry barriers for access to scarce, non-replicable resources, fragmentation, hoarding, pre-emption, market dominance, foreclosure of new entrants, in a context of vertical and horizontal integration, thus creating a potentially harmful situation with no remedies.

Looking at such exclusive property rights, whatever their harmonization and standardization context, many experts question their ability to foster competition and efficiency in spectrum usage on the grounds that strategic use and significant market power often lurk around the corner. They dispute the view held by the FCC and Ofcom that, should significant market power situations arise, they could be dealt with through standard generic competition monitoring rules and procedures. They argue that exclusive property rights on limited resources intrinsically build barriers to entry and have a negative impact on neutrality.

On the contrary, some estimate that exclusive rights are a fair counterpart to heavy financial risks which operators accept when paying their license or investing in a new network. An interpretation of the obligations formally or informally imposed on operators by regulators to subcontract part of their capacity to MVNOs is precisely that they have to mitigate the exclusive character of their license by accepting some kind of sharing.

Property rights with easement

The introduction of easement to exclusive property rights can be justified by recent advances in low power density modulations and dynamic frequency selection systems (DFS), also called dynamic spectrum access networks. Using them, spectrum efficiency can be increased if spectrum exclusive property rights are limited by easement, permitting sharing, overlay and underlay without harmful interference.

Once more, spectrum cannot be seen as a "thing", a ready-made "object" which can be bought, owned, sold. Spectrum resource depends directly on regulations which give a shape and performance to a natural phenomenon, the electromagnetic waves. This means that "exclusive property rights" depend directly on regulation provisions. As an example a band can be fully shared between terrestrial and satellite fixed regulatory services. If appropriate basic technical regulations are set and respected (see Chapter 7, section 7.4), both terrestrial and fixed satellite services may use the resource as they would have full "property" or availability rights on it. Regulation can probably do the same for underlay or DFS terminals in the mobile domain.

In no case can an "exclusive property right" on the spectrum mean that one operator alone can use a frequency band at any time, any place and without any constraint. Easement is a universal rule, at a degree which is regulation dependent.

Collective use

Collective use refers to free access to license exempt bands. In such bands interference can be statistically limited but not fully prevented. However, the benefits are important:

– low entry barriers and certainty of obtaining access;

– quickly addressed niche applications;

– lower demand for licensed spectrum;

– innovation opportunities (anti-monopoly) and freedom;

– possibility of private commons or experimental commons.

As drawbacks, collective use means more significant technical restrictions and higher risks of interference.

Studies are available on collective use, which provide an up-to-date assessment of the potential extent of this category of usage rights[11]. See also Chapter 14, section 14.2.

Deciding on spectrum usage rights

To decide on spectrum usage rights, two levels of analysis have to be considered.

The first is relevant to the institutional acceptability of the property rights framework in the radio spectrum domain. It actually confronts decision makers with difficult social and political choices, very much related to the general institutional setting and mood in each country: as an example, the extension of property rights meets less resistance in the UK and USA than in other countries. A deeper analysis of the legal and economical aspects of spectrum property rights is needed to achieve common views, notably in Europe.

A second level of analysis concerns the technology level, with the reality or feasibility of "flexible technologies" which can justify easement of property rights and permit collective use. Two simple scenarios may be envisaged:

– scenario 1: "flexible technologies" work;

– scenario 2: "flexible technologies" do not work.

Two possible strategies follow:

– allow for the possible advent and significant extension of "flexible technologies" and then either get the greatest possible benefits if they work, or be left with an awkward and inefficient regulation framework if they do not;

– stick to the best known pure property rights, which is the correct decision if "flexible technologies" do not work, but a significant opportunity will be missed if they do.

Those scenarios, strategies and outcomes are summarized in the simple table below.

---

11 *Study on Legal, Economic and Technical Aspects of Collective Use of Spectrum in the European Community* by Mott MacDonald Ltd, Aegis Systems Limited, IDATE, Indepen Ltd and Wik Consult.

	Scenario 1	Scenario 2
Strategies and scenarios for the decade ahead	Advent of flexible use technologies	No significant difference with present technology
Exclusive property rights strategy	Consumer welfare LOW	Consumer welfare MEDIUM
Policy-mix strategy (easing and collective use)	Consumer welfare HIGH	Consumer welfare LOW

The assumptions made here are that:

– sticking to exclusive property rights if flexible technologies are feasible leads to an inferior outcome in terms of consumer welfare;

– flexible technologies are better accommodated by a regime-mix strategy that would include easing property rights and collective use.

If scenarios could be assigned probabilities, the appropriate outcome would depend on them, according to game theory. No objective probability being available, our "best knowledge" can significantly enlighten the decision process. Basically, if the "advent of flexible technologies" scenario is considered probable, the "policy-mix" strategy wins and provides the highest benefit to customers. On the contrary, if this advent is considered improbable, the "exclusive property rights" strategy wins. This analysis is sufficient to satisfy research on flexible use technologies and to encourage the ITU and regulators to study appropriate adaptations of the Regulations.

Step 4: what kind of spectrum assignment mode should be adopted?

Administrative assignment procedures and hybrid procedures

Regulators have made significant efforts over the last decade to keep up with market changes and innovations. Some contend, however, that pure administrative methods, notably formal "beauty contests", have reached a limit and that the situation and inconsistencies in spectrum management regimes actually hinder the deployment of new technologies and services. It can be thought also that regulators are going too far when designing detailed application schemes and objectives for market-oriented business. They may be wrong when they decide "what the market is waiting for".

Advocates of administered neutrality, intended as neutrality achieved within an administered setting, argue that the complexity of spectrum usage, namely the prevention of interference, requires strong control mechanisms. They add that even if it represents some costs, these are smaller than those induced by the multiple conflicts and litigations that would occur in their absence. There should be a trade-off between the costs induced by sub-optimal administrative procedures and the technical monitoring, adjustment and litigation costs they help to avoid.

Hybrid procedures (administrative with bidding as a criterion among others) have the positive effect of combining an administered policy with market-oriented mechanisms, but they may be ambiguous.

The main objectives of administrative procedures are more or less clearly presented:

– welfare considerations with many different aspects to be satisfied;

– domestic and international security concerns (keeping a tight control on spectrum resource);

– preserving national or regional champions (to a limited degree);

– capturing financial rents (possibly by maintaining them?).

Some governments, but not all, are willing to implement this set of criteria and design their procedures accordingly. Others favor a more radical market approach across the board.

Auctions/trading

Whether market mechanisms in the form of auctions and trading can bring competition and efficiency in spectrum usage is hotly debated. Some agencies are optimistic that competition will evidently provide benefits. The pro-market line of reasoning chosen by the FCC Task Force 2002 and Radiocommunications Agency reports 2002 by Martin Cave, refer to the respectable consideration that the introduction of market mechanisms optimizes the usage of spectrum, as of any other resource. It would basically seem that the same faith holds for the policy directions upheld by the European Commission: they propose an extension to the radio spectrum of the general internal market principles, which are a backbone of the EU economic propositions. Auctions and trading are thought to be a way either to correct initial flaws in allocation or assignment, or to allow for changes over time.

However, the existence of a budgetary bias is a case which has been met notably in the UMTS license award procedure: in situations of doubt as to the most efficient spectrum management regime, budgetary considerations may prevail to choose assignment methods providing the maximum income for the government budget. Governments have often clashed with regulators as the consideration of cash-strapped budgets has overwhelmed any other, including consumer welfare. A comparison of economic merits and drawbacks may be sketched, as seen from the regulator point of view.

Deciding on spectrum assignment modes

	Administrative assignment/hybrid	Auctions/trading
Price expectation	−	+
Predation risk	+	−
Entry barriers	+	−
Asymmetry of information on strategy and objectives	−	+
Efficiency	−	+
Winner economic profile	−	+

The evaluations in this table are presented for illustration purposes only. However, it can be estimated that the economic benefits are greater with a market-oriented procedure, and "political" benefits higher with an administrative one.

15.4. Navigating the nine spectrum management regimes: migrations and transitions

It is too early to provide a fact-based assessment of each of the nine spectrum management regimes presented. It is also true that most agencies in charge of the spectrum, even those advocating a strong market orientation, like Ofcom or the FCC, tend to adopt a careful and progressive approach to changes in management methods. The road to a competitive spectrum market is more evolutionary than revolutionary. The specifications and restrictions attached to the frequency usage plan are progressively softened or lifted case-by-case. These changes may lead later to a comprehensive new direction framework. In this sense, the variety of spectrum regimes presented in this chapter can provide a map on which an evolutionary path from one spectrum regime to another may be traced, alongside evolutions in wireless technologies and spectrum usage. The actual implementation of cognitive radio, dynamic frequency selection and underlay uses, for instance, could warrant an extension of spectrum rights easing over time.

Given the strong interaction between decisions on frequency band allocations, services, applications and technology assessments which constitute the basic foundations for the choice of a spectrum management regime, it is advisable not to consider a unique case but to make appropriate choices for relevant sets of bands, services, and technology areas, which may be called wireless clusters. Such clusters would include, for instance, customer personal services with mobile or distributed fixed wireless access, low power and short distance collective use, public security, fixed professional infrastructures, etc. with variants. From this perspective, there might be migrations over time from one spectrum regime to another, to implement changes required by demonstrated positive technological changes and market needs. Instead of a war of doctrines between command and control, market and commons, we could have, when required for a cluster of wireless services, an evolutionary process over time between progressive spectrum regimes. A cross analysis with ITU regulatory services should be maintained in order to maintain coherence with a wider policy for protection against interference throughout the spectrum.

It should be remembered throughout this evolutionary process, however, that, as at a higher level of policy considerations, efficiency depends on pragmatic tough choices but coherence is the key to success. A long-term strategy does not necessarily imply market mechanisms or more social models *per se*, but it should avoid distortions resulting from their application being only partial and erratic: a significant reason for our unsatisfactory situation is the fact that an important part of the commercial spectrum is managed in uncommercial conditions. Convergence will make this long-term coherence necessary (see Chapter 14, section 14.1).

Legacy and institutional factors might play a greater role than markets and technologies in deciding upon immediate and short-term policies. What can be decided if contradictory opinions are expressed on market induced neutrality? What could be a second-best approach if this vision cannot be extended right now across the European Union? One first element of an answer is progressive easement with sharing by collective use and dynamic access. Some sharing methods will necessarily have to take place due to the expected limitations of spectrum available (as was done for WLANs). Furthermore, new technologies such as software-defined systems will help bridge the gap between static (short-term) and dynamic (long-term) efficiency as embedded in the standardization versus non-standardization dilemma. They will substitute families of contiguous technologies to single technologies at any time. They will also facilitate smooth evolutionary transitions over time, enabling dynamic efficiency and lifting the curse of necessary disruptions between discrete standards. The same evolutionary vision could apply to spectrum management regimes.

In this respect, it seems that a widely acceptable migration path towards an efficient model of spectrum management, composed of appropriate regimes for diverse application services, evolving over time, would be the following:

1) preserving some kind of pre-eminent regulation by the States seems widely accepted with a close concert of parties to harmonize their views and decisions;

2) some middle or long-term broader or looser planning (allocation) framework, compliant to international and regional agreements should be designed, taking into account incentive convergence perspectives;

3) a strong protection against interferences must be maintained but with degrees, according to application clusters. Frequency rights should be mitigated accordingly;

4) a comprehensive study of legal, economical and technical aspects related to the "spectrum property" or spectrum utilization rights should be made to build a reference framework for future evolutions and notably to adjust easing provisions;

5) as much collective use and sharing is offered as made possible by technologies without harmful interferences;

6) an incentive international policy must encourage research and experiments to accommodate for the evolution of technologies towards more neutrality;

7) mixed government-industry-led harmonization of frequencies and industry-led standardization of technologies are only made compulsory when needed for strategic reasons;

8) auctions, semi-auctions and trading are carried out when appropriate with a global view that business-oriented services should be managed by market-oriented procedures.

Chapter 16

The Future of the Spectrum:
A Road Towards More Flexibility

The radio spectrum does not exist by itself and spectrum management does not organize an existing and limited natural resource. The spectrum resource is actually created by the regulation and management of a natural and universal phenomenon: radio waves. The ever increasing development of radiocommunications depends on technological progress, regulation and management principles. Together they are mapping the future of the spectrum; they are not independent:

– regulation and management adapt themselves to the possibilities of the present technology;

– technology promises to inspire new ideas for regulation and management.

The spectrum efficiency depends on their convergence on a common strategy and a new "paradigm". Today, the leading idea to improve this global efficiency and "create" resources is flexibility.

However, the spectrum cannot be ruled in a single way. So many different services and applications share the resource that a careful case-by-case analysis is needed to decide which the appropriate direction is.

Among existing services and usages, many find the present situation convenient, particularly if they are internationally managed. This is the case for aeronautical and maritime services, satellite services, scientific research, HF band services and some others. Traditional spectrum sharing methods seem to be well adapted to such needs which want to be protected by a strict regulation, from the beginning.

Defense and public security services probably have similar views, being managed by governmental bodies, such as ministries, which want to be sure of their communication means and cannot accept any dependence on a too mobile and uncertain radio environment. Even when they benefit a "primary" status, they remain very cautious towards the "secondary" services which they allow to share their bands.

If changes should happen for such usages, they will be decided by their users, under their full responsibility.

Many more aggressive questions are asked about civil telecommunications and broadcasting services. Altogether they may only represent 40% of the spectrum use (see Chapter 6), but they represent by far the greatest share of the market activity, notably for private customers and business. This is also the domain where the economic competition is the highest and consequently where the pressure to obtain spectrum resources culminates. Thus, for years now, a fierce debate has been raging to create new opportunities through a deep reform of the regulation: the challenge is to increase the efficiency by greater flexibility.

It is interesting to mention that such bands and uses are now generally managed by independent regulators whose views and policy may depend on their country and legal responsibilities. They try to harmonize their choices, being fully in line with regional conference proposals, such as CEPT in Europe. However, they are criticized by some economic and political actors as being too conservative and technically oriented, with incompatible decisions and short-term strategies. The European Commission clearly wants to change this situation, to relax technical regulations, to open up the market to greater efficiency and to make the economic actors less dependent on the national regulators.

The "3G story" was crucial for such an evolution of minds.

"3G is dead, long live 4G". Some telecommunications observers at the 3GSM trade show in Barcelona, in February 2007, said 3G has failed to deliver on its promises of rapid growth, as a spring board for new mobile multimedia services. Operators worldwide have spent billions buying licenses on spectrum to run 3G networks, only to find that the technology was harder to implement than they had expected. Even where 3G networks are up and running, after years of delays, demand for the video and multimedia services they make possible has been rather disappointing. Expectations have been scaled down to the point where many mobile operators now view 3G as a way to boost their capacity for voice calls in overloaded parts of their 2G networks, rather than as the revenue goldmine once promised.

If it is remembered that 3G was the most heralded project in 2000 with operators, regulators, governments, experts and working parties convinced that it was strategic, it can be thought that technical and business perspectives may sometimes be wrong and damaging. It can also be said that spectrum management should probably not be carried out this way, with tremendous efforts to allocate spectrum to applications which do not meet their market.

Thus, what can be done?

In the spectrum sections of its "proposed changes"[1] to the Review of the European Union regulatory framework for electronic communications networks and services, the European Commission establishes a coherent, comprehensive and original set of forward-looking spectrum policy principles. By emphasizing the role of trading and market flexibility, technology and service neutrality, it departs from traditional spectrum management bases. However, by stressing the need for a clear justification of exclusive usage rights, it differentiates itself from the simple idea that only a free open spectrum market fits all situations.

Let us focus on what should be clarified and what roadblocks must be circumvented for this framework to be properly implemented and provide the positive results that can be expected.

Only a minority of EU Member States have already embraced the flexibility through market perspective. Most of them still think that three issues justify preserving the status quo or only minor departures from it:

– the prevention of interferences;

– the prevalence of harmonization and standardization;

– the institutions.

Today, those issues are translated into general policy principles, in conjunction with critical industrial choices. However, they do not object to more long-term technological evolutions which will impact on spectrum usage and management.

To begin with interferences, the Commission makes it clear that:

> [It is appropriate] to require that the granting of exclusive usage rights on the basis of individual licenses [must] be subject to clear justification that the risk of harmful interference cannot be managed in another way.

---

1 Brussels, 29.6.2006 COM(2006) 334 final, Communication from the Commission to the Council, the European Parliament, the European Economic and Social Committee and the Committee of the Regions, on the Review of the EU Regulatory Framework for electronic communications networks and services.

A problem is that a clear and comprehensive view of interference prevention, what is possible and what is not, based on existing and future probable technologies, is still lacking. Status quo proponents argue that a structured frequency allocation is needed as long as no wide-ranging tested alternative has been demonstrated. This case will remain a major roadblock obstructing the concrete application of the framework and confining it to abstract management principles.

Let us now consider harmonization and standardization issues. The principle of technological neutrality is stressed in the European "proposed changes". Some exceptions can be made to avoid harmful interferences or when they can be strictly justified on the basis of a limited number of legitimate general interest objectives". The words harmonization and standardization do not even appear.

In the past, however, harmonization and standardization have been at the root of the major successes of EU initiatives. Many in the industry and EU governments consider that the road should not be closed to such initiatives in the future. The absence of reference to such possibilities appears to be an error to those who think that it is in the interest of the European industry to harmonize and standardize. The Commission's proposals regarding the spectrum appears, in this respect, to be a complete reversal in attitude, which neglects what has made the European industry successful.

A reasonable policy would be to identify, even restrictively, application areas which deserve harmonization and standardization and, having achieved this, leave the rest to technological neutrality.

Do not let us forget that difficult institutional issues also constantly confront the Commission. A notable example is given by the present impossibility of achieving common views on convergence between audiovisual and telecommunications industry and also their dedicated regulators, which may slow down new services, such as TV on mobile, or, on the other hand, waste spectrum in parallel incoherent projects.

Both audiovisual and telecom industries have views on the hundreds of millions of paying customers to be reached by TV on mobile terminals and on the revenue stream that they may generate. However, the telecom industry feels short of convenient spectrum resources in that respect, as the TV industry can take advantage of frequencies which are made available by digital broadcasting modulations. Such a situation created by the historical spectrum regulation may be found unfair. However, simultaneously, widebands are open to 3G and 4G mobile telecommunications services with uncertainties on their market, as discussed above.

This leads us to definitely long-term considerations. The excessively repeated invocation of spectrum "scarcity" actually reflects three sets of limiting factors which should be overcome.

The first is the technical ability to use spectrum efficiently. Technologies have made enormous progress in recent years, driven both by the economic necessity to extend the spectrum commercial usage and by the hugely increased data transmission and integration needs of the military. Spectrum sharing is now commonly used with more and more sophisticated tools to prevent interferences, such as dynamic access (in GSM for instance). It will expand, as will software-defined and cognitive radios, culminating 10-15 years from now in an extension of dynamic access to ever larger bands and across technologies and services. Technical limiting factors will never completely disappear, but spectrum efficiency should be greatly improved.

Secondly, administrative and economic regulations must also evolve. How spectrum usage rights, in the form of property rights or others, are assigned and handled should be reviewed, depending on usage sectors, to become more efficient. It appears, and this lesson has been learnt at a high price, that market mechanisms, under the watch of public interest, public goods care and competition monitoring, are often more clever than administrative mechanisms. Competition monitoring includes provisions that creates barriers to entry, which though impossible to surmount in all cases, are not created by the spectrum regulation and assignment process themselves. Economically driven players will have an incentive to use spectrum efficiently, unless institutions allow them to do otherwise. It is a common experience that once exclusive property rights are established, they are all the more harshly defended at an institutional level that they are not economically justified. Consequently, and as a matter of efficiency and realism, every time that a "sharing" or "open spectrum" solution seems technically feasible without evident drawbacks, it should be preferred to any other solution.

Like technology, but perhaps at a slower pace, administrative and economic regulations also make progress. Already vast and notable changes have been cleverly handled: spectrum re-farming and relocation operations have been efficiently and quietly carried out. Technical neutrality is no longer ignored in newly issued authorizations. Service neutrality is making a cautiously watched appearance while trading is being progressively introduced. Spectrum institutions, however, are anxious not to disrupt the delicate functioning of the huge and complex operation of multiple wireless devices and services. Some spectrum agencies are more optimistic than others in their administrative and technical ability to navigate on the open seas, once the Pandora's box of market winds has been opened. However, the point is that a process has been launched and will undoubtedly gain momentum as the outcome of the whole set of "flexibility" tools being implemented is observed and assessed.

The next steps of spectrum management evolution will be determined not by ideology or reference to theoretical principles, but by experience.

Thirdly and lastly, broader social and institutional factors are to be considered. Let us only cite one example from large spectrum holders. The arguments put forward by terrestrial TV operators to keep their large and almost free spectrum share include the role they play in social and cultural diversity and their general interest obligations. These are strong requirements to be respected. Yet it remains to be seen whether they justify retaining their whole frequency domain in the digital transition. In fact, mainly historical regulations and lobby pressures are heard in the debate, more than rational and economical arguments, to examine the social role and general interest obligations of terrestrial television and, consequently, to define the conditions and extent of its legitimate access to the spectrum. In this process, a global view of the multiple broadcasting channels on TV sets: digital terrestrial, DSL, cable, satellite, would definitely be examined, in parallel.

To conclude, it appears that technical progress in wireless systems will mostly complement some market mechanisms in fostering the efficient use of spectrum, on the condition that institutional factors do not interfere (barriers to entry). In this sense, there is a good chance that the curse of spectrum scarcity can be lifted or alleviated, without a risk of degrading the current quality of radio services.

Concerning the radio services to private or business customers, debates on industrial strategies and public policies with their impact on spectrum usage and management are urgently required. They present intrinsic technical difficulties: which services, which technologies, in which frequency bands? However, furthermore, they take place on another background debate concerning the regulatory institutions in Europe and their principles for action. In this context a consensus cannot be immediately reached on future methods for the use of the radio spectrum and the appropriate management of the various frequency bands.

There has been a longstanding tradition of technical consultations on spectrum usage among government agencies and interested parties, especially within the ITU and CEPT. Since the UMTS license awarding process, there has also been a growing awareness of the shortcomings of the case-by-case, nation-by-nation approach to spectrum policy and management. CEPT, the European Commission, RSC and RSPG have now been exploring for some time possible future policies with the Commission cautiously making proposals for more coherence between spectrum policy, the internal market, the regulatory framework and competition policy. However, the risk exists of a clash between institutions and cultures, both within and across member states in the EU. The question is: what are the possible outcomes, what are the chances of coming up with a completely new and improved spectrum regime from the top?

Hopefully, in spite of these tensions and in parallel with the progress and convergence of technologies, the future is awakening in the field: frequency bands are used in a more general purpose way, a greater freedom is given for the choice of technologies, even of services. It is the emergence of a new paradigm in which flexibility is the rule and where harmonization and standardization are introduced on a case-by-case basis in view of considerations of markets and industrial situations. This paradigm of increased flexibility can include a mix of market mechanisms and institutional guidance. The environment of the radio operator in Europe is going to change, but in a more evolutionary than revolutionary way. In the interest of industry and consumers, it would be important that this evolution is made in a coordinated way throughout Europe and that the fragmentation of the market is avoided.

Glossary

ALOHA
Collision protocol for radio multiple access.

AMPS: Advanced Mobile Phone System.
First cellular analog radiotelephone network in the USA, late 1970s.

ANFR: Agence nationale des fréquences.
Spectrum management office in France.

APT: Asia-Pacific Telecommunity.
Forum for cooperation on telecommunications matters between Asia and Pacific countries.

ARCEP: Autorité de Régulation des Communications Electroniques et des Postes.
French regulator for electronic communications, notably radiocommunications.

BLUETOOTH
Radio protocol for local communication between terminals.

BNetzA: BundesNetzAgentur.
German regulator, notably for telecommunications.

BR: Radiocommunication Bureau.
Bureau of the ITU radiocommunication sector in charge of implementing regulatory procedures.

BRIFIC: BR International Frequency Information Circular.
International periodical information document issued by the Radiocommunication Bureau according to the frequency regulatory procedures.

CB: Citizen Band.
Customer radio service for personal short distance communications.

CDMA: Code Division Multiple Access.
Multiple access of radio stations to a common frequency band with orthogonal codes.

CEPT: Conférence européenne des postes et télécommunications.
European cooperation organization, notably to harmonize radiocommunications.

CITEL: Comision Interamericana de Telecommunicaciones.
Forum for cooperation on telecommunications matters between American countries.

C/N: Carrier over noise ratio.
A basic ratio to assess the performance of radio links (see Chapter 2).
Also **C/I**: Carrier over interference noise ratio.

COFDM: Coded Orthogonal Frequency Division Multiplex (see OFDM).

CPG: Conference Preparatory Group.
CEPT working group in charge of preparing European common proposals for ITU conferences.

CSA: Conseil supérieur de l'audiovisuel.
Broadcasting regulator in France.

D2MAC
Improved coding standard for analog TV (MAC: multiplex analog component).

DAB: Digital Audio Broadcasting.
Digital audio broadcasting standard. Variants: TDAB for terrestrial broadcasting, SDAB for satellite broadcasting.

dB: Decibel.
Logarithmic ratio. Different logarithmic units derive from this: dBW, dBm, dBV/m (see Chapter 1).

DCS1800
GSM standard adapted to the 1,800 MHz band.

DECT: Digital European Cordless Telephone.
European standard for digital cordless telephone.

DFS: Dynamic Frequency Selection.
Automatic selection of a particular frequency channel, depending on radio environment.

DME: Distance Measuring Equipment.
Equipment to evaluate the distance of a plane to an airport.

DRM: Digital Radio Mondiale.
Digital modulation adapted to audio broadcasting to replace amplitude modulation in HF bands.

DSSS: Direct Sequence Spectrum Spreading.
Protocol to spread the spectrum by logical data extension with orthogonal keys.

DVB: Digital Video Broadcasting.
Broadcasting standard family for digital TV. Variants are DVBT for terrestrial broadcasting, DVBS for satellite, etc.

ECC: Electronic Communication Committee.
CEPT working party to harmonize electronic communications.

ECP: European Common Proposal.
Common proposal for European countries to ITU issued by CEPT.

EFIS: ERO Frequency Information System.
ERO information system on frequency allocations in Europe.

EHF: Extremely High Frequency.
Frequencies between 30 and 300 GHz.

EIRP: Equivalent Isotropic Radiated Power.
Power of an isotropic transmitter which would create a radio field equivalent to the considered station (see Chapter 1).

ERMES: European Radio Message System.
European harmonized paging service.

ERO: European Radiocommunications Office.
Technical permanent bureau of CEPT in Copenhagen.

ETHERNET:
Collision multiple access protocol for wired networks.

ETSI: European Telecommunications Standard Institute.

EUTELSAT
European operator for satellite communication services.

FCC: Federal Communication Commission.
US federal regulator for private and commercial electronic communications.

FDMA: Frequency Division Multiple Access.
Multiple access of radio stations to a common frequency band with channel division.

FHSS: Frequency Hopping Spectrum Spreading.
Protocol to spread the spectrum using fast frequency changes.

FM: Frequency Modulation.
Commonly used to designate the audio broadcasting service using frequency modulation.

FPLMTS: Future Public Land Mobile Telephone System.
Used to designate the future third generation cellular radiotelephone system at the beginning of 1990s. Replaced later by IMT2000.

GALILEO
European project for a worldwide satellite system for radiolocation.

GEO: Geostationary Earth Orbit.
Geostationary satellite orbit. Geostationary satellite.

GLOBALSTAR
Project for a non-geosatellite network for mobile telephony.

GLONASS
Worldwide Russian satellite network for radiolocation.

GMDSS: Global Maritime Distress and Security System.
Worldwide system for security at sea managed according to IMO standards.

GPS: Global Positioning System.
Worldwide American satellite network for radiolocation.

GSM 900 – GSM 1800: "Groupe spécial mobiles" or Global System for Mobile communications.
First digital cellular radiotelephone network (Europe, 1992). Two versions in 900 MHz and 1,800 MHz (DCS 1800).

HAPS: High Altitude Platforms.
High altitude airships which can be used for radiocommunication purposes.

HDFS: High Density Fixed Service.
Radiocommunication fixed service to connect a large number of customers whose premises are not located in the service area from the beginning.

HF: High Frequencies.
Frequencies between 3 and 30 MHz.
Different higher frequency bands are designated HF, VHF, UHF, etc. (see Chapter 1).

HIPERLAN
Local area radiocommunication network standard for high speed data communications.

ICAO: International Civil Aviation Organization.
International organization for the definition of standards, rules and procedures concerning civil aviation.

ICO: Inclined Circular Orbit.
Particular MEO orbit.

IEEE: Institute for Electrical and Electronics Engineers.
American standards institute. Typical standard families for radio networks IEEE802.11, IEEE802.16.

IFRB: International Frequency Registration Bureau.
Former ITU office in charge of filing the international frequency assignments. Replaced by BR.

ILS: Instrument Landing System (or Low-approach System).
Aeronautical equipment to guide planes when landing.

IMO: International Maritime Organization.
International organization for the definition of rules and procedures concerning maritime activities.

IMT2000: International Mobile Telephone 2000.
Radiotelephone standard family for third generation networks.

INMARSAT
International operator for satellite mobile communications, notably for maritime communications.

INTELSAT
International operator for satellite communications, notably for fixed services. A pioneer of the satellite services.

IP: Internet Protocol.

IRIDIUM
Worldwide mobile communication satellite network, using non-geosatellites.

ISM: Industrial, Scientific and Medical.
Equipment using radio waves for purposes other than telecommunications.

ITU: International Telecommunication Union.
International United Nations organization for the regulation of telecommunications, notably radio communications.

LAN: Local Area Network.
Local network for short distance communications.

LEO: Low Earth Orbit.
Low altitude orbit, typically about 1,000 kilometers. Satellite on such an orbit.

LPD: Low Power Device.
Small piece of radio equipment for very short distance communications such as telealarms, telecommands, RLANs, etc. Also SRD.

MEO: Medium Earth Orbit.
Medium altitude orbit, typically about 10,000 kilometers. Satellite on such an orbit.

MIFR: Master International Frequency Register.
Official international file, managed by the ITU, for frequency assignment registrations.

MIMO: Multiple Input, Multiple Output.
Radiocommunication network where different transmitting and receiving stations are combined to improve the signal reliability (digital diversity).

MMDS: Multichannel, Multipoint Distribution System.
TV broadcasting for local coverage using semi-directive radio beams. Also MVDS.

MPEG: Motion Picture Expert Group.
TV digital code family. Variants: MPEG2, MPEG4, etc.

NATO: North Atlantic Treaty Organization.
International organization to coordinate the military forces of countries which are Members of the Treaty.

NMT: Nordic Mobile Telephone.
First generation cellular analog radiotelephone network. Scandinavian countries.

NTIA: National Telecommunications and Information Administration.
American administration in charge of governmental communications.

Ofcom: Office of Communications.
British regulator for electronic communications.

OFDM: Orthogonal Frequency Division Multiplex.
Digital radio modulation using orthogonal sets of carriers.

PAR: Puissance Apparente Rayonnée.
Power of a half-wave dipole transmitter which would create a radio field equivalent to the considered station (see Chapter 1).

PFD: Power Flux Density.
Local power flux carried by an electromagnetic wave.

PMR: Private Mobile Radio.
Private radio network, especially for mobile communications.

QAM: Quadrature Amplitude Modulation.
Digital amplitude modulation.

QPSK: Quaternary Phase Shift Keying.
Digital modulation using four-state phase shifts.

RA: Radiocommunications Agency.
Former frequency regulator in the UK.

RegTP: Regulierungsbehörde für Telekommunikation und Post.
Former designation of regulator in Germany (now BNetzA).

RFID: Radio Frequency Identification.

RLAN: Radio Local Area Network.
Short distance digital radio network for local communications, particularly for data.

RPC: Radio Preliminary Conference.
ITU expert preparatory meeting before a World Radiocommunication Conference.

RR: Radio Regulations.
Rules issued by the ITU for radiocommunications.

RRB: Radio Regulations Board.
Expert board of the ITU radiocommunication sector for interpretation of the Regulations.

RRC: Regional Radiocommunication Conference.
ITU conference to adapt Radio Regulations in a particular Region.

R&TTE: Radio and Telecommunications Terminal Equipment.
European directive on terminal market.

SARSAT: Save and Rescue Satellite.
International satellite service for location of rescue calls.

SAR: Specific Absorption Rate.
Radio power which can be absorbed by a living biological tissue without being disturbed.

SDAB: Satellite Digital Audio Broadcasting.
See DAB.

SHF: Super High Frequencies.
Frequencies between 3 and 30 GHz.

SKYBRIDGE
Project of a non-geosatellite network for worldwide customer access to the fixed service.

SOLAS: Safety Of Life At Sea.
International convention for life safety at sea.

SRD: Short Range Device.
Low power equipment for short range communications. Also LPD.

TCAM: Terminal conformity assessment and market surveillance committee.
Consultation committee set up by the European Commission in order to follow the
application of the R&TTE Directive on the marketing of terminals.

TDAB: Terrestrial Digital Audio Broadcasting.
See DAB.

TDMA: Time Division Multiple Access.
Multiple access of radio stations to a common frequency band with time sharing.

TELEDESIC
Project of a non-geosatellite network for worldwide customer access to the fixed
service.

TETRA
European standard for digital radiotelephone trunk network.

TETRAPOL
European standard for digital radiotelephone trunk network.

TFTS: Terrestrial Flight Telephone System.
European mobile telephone network for airline passengers.

UHF: Ultra high frequencies.
Frequencies between 300 and 3,000 MHz.

UMTS: Universal Mobile Telephone System.
European standard for digital third generation mobile telephone cellular networks.

UWB: Ultra wideband.
Radiocommunication system with digital signals spread over very large frequency
bands.

VHF: Very High Frequency.
Frequency between 30 and 300 MHz.

VOR: VHF Omnidirectional Range.
Aeronautical equipment for airplane guidance to airports.

VSAT: Very Small Aperture Terminal.
Satellite network to connect customers to telecommunications services directly with
a small individual antenna.

WAPECS: Wireless Access Policy for Electronic Communication Services.
Concept of digital multi-purpose radio systems.

WCDMA: Wideband CDMA.
Variant of CDMA using wideband modulation.

WiFi
Radio local area network standard for short distance, medium speed, data communications. Also IEEE 802.11b.

WiMAX
Radio local network to connect customer premises to high speed telecommunications networks. Also IEEE 802.16.

WLL: Wireless Local Loop.
Radio local network to connect customer premises to telecommunications networks.

WRC: World Radiocommunication Conference.
ITU world meeting, notably for updating the Radio Regulations.

References

ANFR (Agence Nationale des Fréquences, France)
http://www.anfr.fr/index.php?&page=liens

Bundesnetzagentur (Germany)
http://www.bundesnetzagentur.de

CEPT (Conférence Européenne des Postes et Télécommunications)
http://www.ero.dk

European Commission Spectrum Policy
http://ec.europa.eu/information_society/policy/radio_spectrum/index_en.htm

Federal Communications Commission (USA)
Spectrum Policy Task Force
http://www.fcc.gov/sptf

Ofcom (UK)
http://www.Ofcom.org.uk/radiocomms

Review of Radiocommunications Acts and of the Market-Based Reforms and Activities undertaken by the Australian Communications Authority
February 2002
http://www.pc.gov.au/inquiry/radiocomms

Review of Radio Spectrum Management
An Independent Review for the Department of Trade and Industry and HM Treasury
by Professor Martin Cave
March 2002
http://www.spectrumreview.radio.gov.uk

SPORTVIEWS research project
http://www.sportviews.org

ITU (International Telecommunication Union)
http://www.itu.org

Index